IMAGES OF ASIA
Series Adviser: SYLVIA FRASER-LU

The Birds of Sumatra and Kalimantan

Titles in the series

Schneider's Pitta, p. 41.

The Birds of Sumatra and Kalimantan

DEREK HOLMES

Illustrated by
STEPHEN NASH

SINGAPORE
OXFORD UNIVERSITY PRESS
OXFORD NEW YORK
1990

Oxford University Press

Oxford New York Toronto
Delhi Bombay Calcutta Madras Karachi
Petaling Jaya Singapore Hong Kong Tokyo
Nairobi Dar es Salaam Cape Town
Melbourne Auckland
and associated companies in
Berlin Ibadan

Oxford is a trade mark of Oxford University Press

ISBN 0 19 588971 1

Printed in Singapore by Peter Chong Printers Sdn. Bhd.
Published by Oxford University Press Pte. Ltd.,
Unit 221, Ubi Avenue 4, Singapore 1440

Acknowledgements

THE concept of popular books on Indonesian birds, formulated by
Bill Harvey and myself in 1976, reached fruition only in 1989, with
the publication of the companion volume on the birds of Java and
Bali. The delay resulted from the earlier lack of an artist who was
familiar with the birds in the field, and I am extremely grateful to
Stephen Nash who so enthusiastically agreed to undertake this task
on top of a schedule that was already heavy.

Appreciation is offered especially to those Indonesian members of
the Indonesian Ornithological Society whose moral encouragement
and friendship will always be valued, and to the many field ornithologists who share their knowledge in discussion and correspondence.
Finally, I wish to thank Anne Nash for her time-consuming work in
editing and typing the text and her assistance in proof-reading.

Jakarta   D. A. H.
April 1990

Contents

Introduction

ANYONE who has an interest in the birds of the Indonesian forests encounters a problem. The variety of birds is very wide, but the field guides or handbooks available are usually comprehensive and confusing or daunting to the layman. The problem is compounded for the overseas visitor who is not yet familiar with the bird families, which is the first step in identification. The author has long felt the need for popular bird books in Indonesia, such as those in his home country which first inspired his own enthusiasm in his youth. The present book attempts to fulfil this need, by providing a simple introduction to the full range of birds found in Sumatra and Kalimantan.

One or more representative species are described from nearly every bird family found on the two islands. The species have been selected carefully so that the birds described are those which a layman is most likely to find, or to be able to identify, on visits to the different habitats. The main descriptions cover 148 species, of which 133 are illustrated in colour and the rest in black and white. Reference is made in the text to a further 129 species in order to illustrate the range and variety of birds in each family and guide the reader towards their identification.

The name 'Kalimantan' is used to refer to the Indonesian portion of Borneo, which occupies about 70 per cent of the island. However, Kalimantan has the same avifauna as the rest of Borneo, so this book applies to the entire island, although there are a few Bornean species that have not yet been reported from Kalimantan (either because they are confined to the higher mountains of northern Borneo, or because the Indonesian part of the island has been less thoroughly surveyed).

It would be impossible, and counter-productive, to describe every bird of such a rich area in a popular book. Sumatra has no less than 600 species, of which about 450 are resident. Borneo has only slightly fewer, and most are the same species (a careful distinction is made in the text for birds that are not common to both islands). At various periods of the Ice Age, Sumatra and Borneo formed part of a larger land-mass which also included Java and Bali, the Malay Peninsula and southernmost Indo-China, and is known as the Sundanese sub-continent. This region shares a common avifauna, and it is only changes in sea-level, the joining and separation of the different land areas, and climatic changes and habitat development that have allowed variations in the avifauna to arise among the different island components of the sub-continent.

Consequently, the number of endemics (species confined to one island or region) is quite low. Sumatra has only 17 endemic species, of which 4 are confined to its western off-shore islands and most of the remainder are montane. Borneo has a larger total of 33 endemics (of which 18 are montane) but 7 of these have not yet been recorded in Kalimantan.

The total number of endemics found in Sumatra and Borneo together is 61 (or 52 in Sumatra and Kalimantan). The Appendix lists all species that are endemic to the Sundanese region of Indonesia (Sumatra, Kalimantan, Java and Bali). These are the species that are not found in bird books covering Peninsular Malaysia, or mainland South-East Asia in general.

Nearly all the author's bird-watching activity in South-East Asia has been in the lowland Sundanese forests, since his first arrival in Brunei in 1968. Bird-watching in the lowland forests is never easy. Apart from the discomfort, it requires skill and a considerable degree of luck. The birds are very difficult to see, and the ornithologist must rely heavily on his knowledge of their calls in order to determine what species are present. For example, the author has never seen an Argus Pheasant in the wild, though it can be heard nearly every day spent in good forest! Birds which do not often call, or whose calls are not known, are very rarely seen. However, knowledge of both the habits and the voices of the forest birds has advanced significantly over the last two decades, and this book gives a lot of emphasis to the description of calls whenever this is an aid to identification.

In the selection of species for coverage, the author freely acknowledges a bias towards the more prominent or colourful resident birds of the forest, as these are the ones which the layman is most likely to wish to identify. The more determined naturalist will be fascinated by the various calls or songs of the smaller birds but the detailed identification of the many similar babblers and bulbuls, spiderhunters and female sunbirds, is beyond the scope of this book. Likewise, the migrant birds are not given prominence here, as they are fully described in the literature available to the expert who needs this. For example, only serious ornithologists are likely to reach the remote and muddy east coast of Sumatra, and they will already have the literature and knowledge required to identify the wading birds that occur there.

There is a wide range of habitats in the region. The mountains provide excellent bird-watching, except in some areas where deforestation is nearly complete. Many of the mountains of Sumatra are quite accessible and much frequented by visiting bird-watchers; those in Kalimantan are very inaccessible, but many discoveries surely await those who have the resources to reach them.

The dipterocarp forests of the lowlands provide the richest habitat, although this is often belied by the total silence that usually confronts the observer in the middle of the day—it is essential to get into the forest at dawn or soon after. There is a great demand for the conversion of lowland forests to agriculture, and it seems inevitable that the richest natural habitat in the Sundanese region will soon become confined to a few areas set aside as reserves for conservation or for managed timber production. Many forest birds are incapable of surviving in any other habitat, and species will eventually become extinct if their remaining natural habitats become too small and scattered.

The wetland forests of the coastal plains are not quite so rich as those on the drylands, but are nevertheless very valuable; however, difficulties of access may deter all but the most enthusiastic observer. The importance of the more open wetlands—the open lakes and marshes and the muddy coasts—has been overlooked until recently, and some exciting discoveries of waterbird habitats have been made since. Only lately, for example, has the east coast of Sumatra been discovered to have global importance as a wintering and transit ground for waders. Unfortunately, the bird life of many of these wetlands, such as the *lebak* of South Sumatra, or the swamps and lakes of the Kapuas, Barito, and Mahakam rivers of Kalimantan, is severely threatened by human activity.

It is the objective of this book to encourage a greater interest in birds in particular, and wildlife in general, and to foster an increased awareness of the need to protect an adequate area of the richest natural habitats in the world while there is still time. To this end, it is planned to produce an Indonesian-language version, so that the book will be more widely accessible to those who have the ultimate responsibility for conserving such a rich and fragile heritage.

The plates and figures, which have all been prepared by Stephen Nash, illustrate the male bird in every case (except Raffles' Malkoha). They have not been drawn to scale, as there is often a range of different-sized birds on each plate, but the size of each species is given in the text. The scientific name is given for every species, which is essential even in a popular book because English names vary between authors; indeed opinion on scientific nomenclature also varies, as may be noticed by readers of the companion volume on Java and Bali. Indonesian names are not given, as these are still being compiled by the Indonesian Ornithological Society; it is a massive undertaking to find suitable names for the more than 1,500 species that comprise the Indonesian avifauna.

Cormorants

ORIENTAL DARTER
Anhinga melanogaster (90 cm). Sumatra and Kalimantan.   Figure 1

The darter, sometimes known as 'Snakebird' because of its dispro-
portionally long neck, is the only member of the cormorant
family that occurs quite widely in Sumatra and Kalimantan,
though in small numbers. It occupies fresh- and brackish-
water swamps of the lowlands, roosting and breeding
in trees and feeding in open water. It is an expert
swimmer and diver, and will sometimes swim
with its body below water and just its head
and upper neck visible, like a periscope. Like
all cormorants, it commonly holds its wings
out to dry when resting in the tree-tops. In flight,
the neck is held outstretched but slightly kinked.
Adults show varying amounts of white in the
black plumage, or a silvery sheen, while
young birds are browner. Darters nest
gregariously in swamp trees and mangroves,
but very few sites have yet been located.

1. Oriental Darter

Herons, Egrets, and Bitterns

These are the tall, long-necked, and long-legged wading birds of the
marshes and wet ricefields, which have long dagger-like bills for
stabbing at fish and crustaceans. In flight, all hold their necks in an
'S'-shape, and fly with rather slow, steady beats on rounded wings.
The herons are the tallest, and the GREAT-BILLED HERON *Ardea
sumatrana* reaches a length of 115 cm; this is a rare and solitary heron
of remote muddy coasts and mangroves. Most often seen is the
PURPLE HERON *Ardea purpurea* (97 cm), a light purple bird of
fresh-water marshes.

The egrets are the graceful, white birds of wet ricefields, muddy river banks, and shallow swamps. There are several species, ranging in size from the GREAT EGRET *Egretta alba* (90 cm) to the more squat JAVAN POND HERON *Ardeola speciosa* (45 cm). The latter is quite a familiar bird; it has cryptic colouring when perched, but reveals pure white wings when it flies. Most of the true egrets are pure white, and they are not always easy to identify.

The bitterns are more secretive birds of fresh-water reed-beds and swamp scrub, and are most often seen when they make short flights above the reed-beds, usually at dawn or dusk. The CINNAMON BITTERN *Ixobrychus cinnamomeus* (40 cm) is quite widely distributed in damp thickets in riceland and riverine areas, and as its name implies, it has a cinnamon colour.

Storks and Ibises

All the storks and ibises are uncommon, and the last populations of some very rare and endangered species are found in Sumatra and Kalimantan. Storks are rather heavy, tall, and gaunt birds with long legs and necks and massive, dagger-like beaks. They fly with necks outstretched, and are experts at soaring in thermals, sometimes to a considerable height. The ibises have a more dumpy build, and long, down-curved bills.

MILKY STORK
Mycteria cinerea (100 cm). Sumatra.   Plate 1b

The Milky Stork is white, except for a dark breast band (not illustrated), and black tail and flight feathers. The bill is very slightly down-curved. This stork feeds on the inter-tidal mud-flats, on mud so soft that a man cannot walk. It walks very deliberately, a few paces at a time, probing its bill deep into the mud. Until recently, this stork was believed to be seriously endangered, but now a population of some 5,000 birds has been located along the muddy east coast of Jambi and South Sumatra. Small parties can also be seen along the east coasts of Aceh and North Sumatra, such as at Lhokseumawe and Belawan. The discovery of this large population, together with other water-birds, is one of the most exciting recent ornithological events in Indonesia.

The LESSER ADJUTANT *Leptoptilos javanicus* (115 cm) is a large and evil-looking stork that can sometimes be seen, usually singly or in twos or threes, stalking rather grotesquely in the open habitats of coasts, estuaries, ricefields, and river valleys in both Sumatra and Kalimantan. Few of its breeding sites are known. It is mainly black above and white beneath, and the bill is very massive. The author has seen up to 13 feeding on the ploughed land of cassava fields in Lampung, unconcerned by the tractors working around them.

STORM'S STORK
Ciconia stormi (90 cm). Sumatra and Kalimantan.   Plate 1a

As the Storm's Stork is nearly extinct in Peninsular Malaysia and South Thailand, its small remaining population in Sumatra and Kalimantan is very important. It is black, with a white neck and lower belly. The black of its breast extends up the front of the neck. The bill, facial skin, and legs are red. It occurs in riverine and swamp forests, feeding in the small seasonal swamps that are found behind the river banks, but it is most often seen either in soaring flight, or perched in twos or threes on trees lining the river banks. A nest discovered in mangroves in South Sumatra in 1989 was only the second known to science.

In Sumatra, there is risk of confusion with the WOOLLY-NECKED STORK *Ciconia episcopus* (90 cm). This stork, which ranges from Africa to South-East Asia, is extremely similar to the Storm's Stork, but it has rather cleaner plumage, with a pure white neck; the bill and facial skin are greyish-black, though there may be a reddish tinge. It prefers more open country, but the habitats of the two species overlap to some extent.

WHITE-SHOULDERED IBIS
Pseudibis davisoni (75 cm). Kalimantan.   Plate 1c

This intriguing bird must be one of the least known and most endangered species in Indonesia. One relict population occurs in Kalimantan, and another in Indo-China. It is a large black ibis, with a bare head, and a white patch in the wings which is said to be conspicuous in flight. The legs are red. There are only a few records from Kalimantan, where it appears to inhabit the banks of the large rivers in the forested interior, presumably feeding in adjacent open swampy areas. Most recent sightings have been along the upper Mahakam River, which appears to be the centre of its population. There is a danger that the bird could become extinct in Kalimantan before it is known how to protect it.

The BLACK-HEADED IBIS *Threskiornis melanocephalus* (75 cm) is a large, rather heavy bird of remote lowland coasts. It is white, with a black neck and large, black, down-curved bill. It occurs from India to South-East Asia, and over 800 have recently been counted on the Jambi and South Sumatra coast, an important centre of its population.

Ducks

Wild ducks are not a very conspicuous part of the Indonesian avifauna, and only the chestnut-brown treeducks (sometimes known as whistling teal) are common. These are locally known as 'belibis', from their frequently uttered whistles, and they are quite common in all fresh-water wetlands. The LESSER TREEDUCK *Dendrocygna javanica* (40 cm) is common in Sumatra, but the WANDERING TREEDUCK *D. arcuata* (58 cm) is more common in Kalimantan, and is distinguished by its larger size and white patches on the flanks and rump. Of the many species of northern duck that migrate south in winter, only the GARGANEY *Anas querquedula* (40 cm) reaches Indonesia with any regularity.

COTTON PYGMY GOOSE
Nettapus coromandelianus (35 cm). Sumatra and
    Kalimantan.   Plate 1d

This attractive little duck is a rather uncommon resident, but has been sighted by the author in northern Lampung, Jambi, and the Barito swamps. The male in breeding plumage is a striking black-and-white bird, with the black parts glossed green. In flight, there is a broad white band in the wings. The female has less strongly contrasting plumage.

WHITE-WINGED WOOD DUCK
Cairina scutulata (75 cm). Sumatra.   Figure 2

This is another of Indonesia's intriguing relict populations that is in danger of becoming extinct. Formerly it ranged from North-east India and Burma south through the Malay Peninsula to Sumatra and Java. It is now extinct in Malaya and Java, and very rare throughout its range. Sumatra possibly holds the largest wild population. All recent records have been from Jambi south to Lampung, where it frequents the backswamps of the large river valleys and probably also the coastal swamps. It feeds in open swamp but breeds in holes in big forest trees. It is thinly distributed, and it is difficult to hazard a guess of the total population. Although the Sumatran bird may tolerate more open conditions and human disturbance than its cousins in continental Asia, serious environmental degradation in this area threatens its continued existence.

2. White-winged Wood Duck

It is a large, goose-like duck, related to the domestic Muscovy Duck. Blackish, but with white wings and neck, Sumatran birds often show variable patches of white elsewhere on the body (partial albinism). It feeds mainly at night, especially when there is a strong moon, and the best time to look for it is at dusk or dawn, when it flies to and from its feeding grounds, uttering a far-carrying call that is a curious mixture of a honk (or croak) and a whistle. One or two can usually be seen along the rivers in Way Kambas.

Birds of Prey

The birds of prey, or raptors, form quite a large and diverse group of predatory birds. All have hooked bills and strong talons, but the methods of hunting are varied, including soaring, hovering, diving, quartering the ground, and swift flight through the canopy.

BRAHMINY KITE
Haliastur indus (45 cm). Sumatra and Kalimantan.   Plate 2c

This chestnut-and-white bird is a familiar resident of rivers, coasts, and harbours, and is generally the identity of the 'sea eagle' so often reported to ornithologists by their friends. It is not an eagle, although it has similarly broad wings and tail, and unlike the true kites, the tail is not forked. The adult shown in Plate 2 is distinctive, but the young bird is dull brown and spotted, and so may be confusing. The call is shrill and quavering.

The WHITE-BELLIED SEA EAGLE *Haliaeetus leucogaster* (70 cm) is a much larger bird of the coasts. Its wedge-shaped tail is distinctive, and the under-parts in the adult are mainly white.

BLACK-SHOULDERED KITE
Elanus caeruleus (33 cm). Sumatra and Kalimantan.   Plate 2b

The Black-shouldered Kite, with its pointed wings, is a graceful raptor that is usually seen hovering over open country. It is pale grey, whitish below, and has a distinctive black patch on the bend of the wing. It is the only resident raptor in Sumatra and Kalimantan that hovers.

CRESTED SERPENT-EAGLE
Spilornis cheela (50 cm). Sumatra and Kalimantan.   Plate 2d

The novice may have difficulty in identifying the several eagle species that occur, but the Crested Serpent-Eagle is the most widely distributed, and readily identified in overhead flight by the single broad white band in the tail and another close to the hind margin of the under wing. The call is a shrill di- or tri-syllabic note which often draws attention to a soaring bird or pair. When the bird is perched, an observer should look for the pronounced crest, although this is not a distinguishing feature.

Among the larger eagles, the two most common species are the BLACK EAGLE *Ictinaetus malayensis* and the CHANGEABLE HAWK-EAGLE *Spizaetus cirrhatus*. Both are 70 cm in length. The Black Eagle appears all black, with yellow feet, and occurs mostly in the mountainous areas. The Changeable Hawk-Eagle is usually seen soaring over lowland forests. The wings are very broad and the tail is long and rounded. The plumage is variable, but most birds are brown above and paler below, and there are fine dark bars on the under wings and tail.

LESSER FISH-EAGLE
Ichthyophaga nana (60 cm). Sumatra and Kalimantan.   Plate 2a

Unlike the Sea Eagles, which have wedge-shaped tails, the Fish-Eagles have rounded tails. The Lesser Fish-Eagle is a bird of small rivers in the forests, and is usually seen flying off into the riverside trees as one's boat approaches. It is a brown bird, grey about the head, with white on the belly, and an all-dark tail.

The GREY-HEADED FISH-EAGLE *I. ichthyaetus* (75 cm) occurs on the larger rivers and open swamps. It is a larger bird with a pale grey head, and a white base to the black tail, which is distinctive in flight. The author has heard this bird make extraordinary loud, raucous calls in the rich open swamps of the Barito region of South Kalimantan.

7

BLACK-THIGHED FALCONET
Microhierax fringillarius (15 cm). Sumatra and
Kalimantan.   Plate 2e

Casual bird-watchers often overlook the falconets, not realizing that
such small birds are falcons, but their ferocity and hooked bills soon
reveal their affinity. The falconet occurs on forest edges and in lightly
wooded country, where it perches openly in low trees, making short,
direct flights. The plumage, mainly black above and white with some
rufous below, is attractive, and in their unobtrusive way, these birds
are quite distinctive.

Game-birds

GREAT ARGUS
Argusianus argus (70–200 cm). Sumatra and Kalimantan.   Plate 3a

Although an observer may never see one, the Argus is familiar
because of its loud calls. These can be heard most days in lowland and
hill forest, and typically consist of a loud, far-carrying 'ku-wao' from
which it gets its local name of 'Kuau'. Another call is a long series of
up to 30 mono- or slightly di-syllabic notes in a series on one pitch,
becoming slower and rising slightly towards the end. Surveyors in
the forest ·sometimes come across the male's dancing grounds, circular clearings of 3 m or more in diameter, from which every twig
and leaf has been removed, which he uses for his dancing displays to
his mate. The length given above includes the tail of the male bird,
which can reach up to 150 cm. The secondary feathers of the male's
wings are likewise enormously lengthened and decorative, in the
manner of ornamental pheasants and peacocks, with beautiful patterns
of rufous, buff, and black.

CRESTED FIREBACK
Lophura ignita (65 cm). Sumatra and Kalimantan.   Plate 3b

The other pheasants of the forest are much less vocal and therefore
less conspicuous. Nobody really knows how common they are, but

those confined to lowland forest must surely be treated as threatened. Both this species, which is shown in Plate 3, and the CRESTLESS FIREBACK *L. erythrophthalma* (50 cm) are lowland pheasants. The latter is distinguished by the lack of a crest, red facial skin, and more rufous tail. There are other pheasant species, however, and also the smaller partridges of the forest; some are endemic to either island, all are little known, and the casual observer is likely to meet them only by accident.

In Sumatra, but not Kalimantan, one may hear quite commonly the RED JUNGLEFOWL *Gallus gallus*, which has a crowing call very like that of the domestic cockerel though with a rather thinner and more abrupt quality.

BLUE-BREASTED QUAIL
Coturnix chinensis (15 cm). Sumatra and Kalimantan.   Plate 3c

A much smaller game-bird, the Blue-breasted Quail is common in open grasslands, dryland crops, and ground scrub. The male is distinctively marked, as shown in Plate 3, but it is very difficult to obtain a good view. Generally they run to hide, or leap up on approach and are seen only as a whirring brown ball diving into cover. The female is dull brown.

Crakes and Rails

Most crakes and rails are secretive birds of the marshes, difficult to observe as they quickly run for cover in the reed-beds or wet thickets when disturbed. They fly weakly, with dangling legs. The short tails are frequently flicked up, usually revealing a striking white, rufous, or barred patch under the tail.

WHITE-BREASTED WATERHEN
Amaurornis phoenicurus (33 cm). Sumatra and Kalimantan.   Plate 4a

This is quite a conspicuous bird that is often seen in a variety of damp locations such as roadside ditches, thickets, and ricefields. As Plate 4 shows, the plumage is distinctive, and it frequently cocks its tail as it walks, revealing the chestnut under tail coverts. The call is a prolonged cacophony of grunts and chuckles that develop into a long series of 'kru-wak, kru-wak' calls, sometimes for minutes on end.

SLATY-BREASTED RAIL
Rallus striatus (25 cm). Sumatra and Kalimantan. Plate 4b

The smaller crakes and rails of the reed-beds are extremely secretive, and a combination of stealth and luck is required to see them, usually in the evening or early morning. The Slaty-breasted Rail is a larger species, but it has the combination of chestnut and finely barred plumage that is typical of the group. It occurs quite commonly also in dry grassland plains.

Jacanas

PHEASANT-TAILED JACANA
Hydrophasianus chirurgus (30 cm). Sumatra and
Kalimantan. Plate 4c

Although not very common, jacanas are very striking, with their long legs and toes which are specialized for walking on floating vegetation in swamps, earning them the name of 'lily-trotters'. They can swim and even dive if required. The flight is weak, and like rails, the legs trail behind, and the wings are short and rounded. The Pheasant-tailed Jacana is a winter visitor from the north. During this season, it is mostly brown above and white below, with a dark breast band and yellow on the hind-neck. In flight, the wings are wholly white. Before they depart north, the males begin to exhibit the beautiful plumage shown in Plate 4, with a tail of 25 cm.

The BRONZE-WINGED JACANA *Metopidius indicus* (28 cm) may be resident in southern Sumatra. This has no white in the wings, and lacks the long tail. However, it is a beautiful, bronzy-coloured bird, with a long white eye-brow.

The common jacana in the Barito swamps of South Kalimantan is the COMB-CRESTED *Irediparra gallinacea* (23 cm), which carries an extraordinary red wattle on the front of the head. This is an Australasian species that just reaches Sulawesi and this south-east corner of Kalimantan.

Plovers and Waders

This group includes all the brown wading birds that are found around the estuaries and tidal mud-flats of the coast and on muddy patches inland, mostly during the northern winter and on migration. They range in size from the stints (15 cm) to the tall greenshanks (35 cm) and long-billed curlews (58 cm), but the identification of the smaller species in particular can be very difficult, and beyond the scope of this book.

Recent surveys have identified the muddy east coast of Sumatra, from the Kuantan to the Musi, as an extremely important transit and wintering area for waders, with over 100,000 birds present. The most abundant species are MONGOLIAN PLOVER *Charadrius mongolus*, WESTERN CURLEW *Numenius arquata*, BLACK-TAILED and BAR-TAILED GODWITS *Limosa limosa* and *L. lapponica*, COMMON REDSHANK *Tringa totanus*, TEREK SANDPIPER *Xenus cinereus*, CURLEW SANDPIPER *Calidris ferruginea*, and ASIAN DO-WITCHER *Limnodromus semipalmatus*. The Dowitcher is an endangered species, and the discovery of up to 13,000 migrant birds in Sumatra is of major international significance.

Inland, the most frequent species are GOLDEN PLOVER *Pluvialis dominica*, COMMON SANDPIPER *Actitis hypoleucos*, and WOOD SANDPIPER *Tringa glareola*. Golden Plovers (25 cm) roost on airfields, football pitches, and golf courses, where their 'tu-ee' calls may often be heard as they come to roost at night. The Common Sandpiper (20 cm) is a common migrant to rivers and shores, but is usually seen singly, while the 'tiss-iss-iss' calls of the Wood Sandpiper (23 cm) are often heard over ricefields.

Terns

Terns are graceful birds of the open seas and, sometimes, inland waters. They have narrow wings, long bills, and rather long, forked tails. They feed by diving into the water, sometimes hovering first (*Sterna* sp.), or by quartering restlessly back and forth across marshy areas and shallow tidal waters, plucking food from the surface but not diving (the marsh terns, *Chlidonias* sp.). About ten species visit these coasts.

BLACK-NAPED TERN
Sterna sumatrana (30 cm). Sumatra and Kalimantan.　　　　　Plate 4d

The Black-naped Tern is distributed all round the coasts, but especially on the rocky shores. It does not congregate in flocks like other terns, nor does it occur inland. It breeds on off-shore rocky islands, and like all terns, will often rest on fishing platforms. It is whiter than most terns, with only a narrow black collar around the back of the nape. This tern has a black bill, and long tail streamers which give the tail a deeply forked appearance.

Pigeons

There are some 20 species of pigeons and doves, but they fall into a number of distinct genera, and identification of most species is not difficult if good views can be obtained. Some are quite colourful, and their gurgling or cooing calls are attractive. Most are forest birds, but being strong fliers, they range quite widely over a variety of habitats.

LITTLE GREEN PIGEON
Treron olax (20 cm). Sumatra and Kalimantan.　　　　　Plate 5a

The green pigeons are plump fruit-eaters of the forest. The females are mostly green, but the males are more diversely coloured and the wings are maroon in many species. The best place to observe them is at fruiting fig trees, where they will congregate greedily together with barbets and hornbills. Plate 5 shows that the male Little Green Pigeon is quite a colourful bird. It frequents the canopy of lowland forest, where the expert can identify it by its call. Most green pigeons have a short chuckling cooing note or gurgle, but in this species it is preceded by a nasal winding-up whine.

Another common lowland species is the THICK-BILLED GREEN PIGEON *Treron curvirostra* (27 cm), the male of which also has the maroon mantle, but is otherwise mainly green, with a greyish cap, pink neck, and red base to the bill.

CINNAMON-HEADED PIGEON
Treron fulvicollis (25 cm). Sumatra and Kalimantan.   Plate 5b

Plate 5 shows that the male of this species is mainly reddish, ranging from chestnut on the foreparts to maroon on the mantle. By contrast, the female is almost entirely green. This pigeon is the commonest of the group in the rather scrubby sand and peat forests of the southern regions of both Sumatra and Kalimantan.

The PINK-NECKED PIGEON *T. vernans* (27 cm) occurs in open country and secondary growth, and is one species in which the male has a green rather than maroon mantle. The foreparts resemble those of the Little Green Pigeon, but it has a broad, pale pink collar and breast, above an orange lower breast band.

GREEN IMPERIAL PIGEON
Ducula aenea (43 cm). Sumatra and Kalimantan.   Plate 5f

The Imperial Pigeons are heavier birds of the forest canopy. The Green Imperial Pigeon is commonly seen in flight across forest clearings, especially in the morning and evening, and its deep, soft 'hroom' call, preceded by a quiet clicking note audible only at close quarters, is one of the characteristic sounds of lowland forest.

It is replaced in the mountains by the MOUNTAIN IMPERIAL PIGEON *D. badia* (47 cm), which is a duller-coloured bird, lacking the green mantle and tail. The call is a booming sound, 'hoo-HOOM', which echoes across the valleys.

LITTLE CUCKOO-DOVE
Macropygia ruficeps (30 cm). Sumatra and Kalimantan.   Plate 5c

The Cuckoo-Doves are rufous-brown, slender doves with long tails. They have swift flight, and frequently all that is seen of them is two or three birds darting above the canopy on the sides of a valley in hilly country. The plumage of this species is mottled rufous, and faintly barred. The call is a rapid 'croo-wuk, croo-wuk' repeated a few times, though at a distance it sounds like 'wuk, wuk, wuk ...'.

In the mountains of Sumatra, it is replaced by the larger BARRED CUCKOO-DOVE *Macropygia unchall* (35 cm), which has a slower, disyllabic 'wo-oo, wo-oo' call, at a rate of about one per second. In hilly country in Kalimantan, the BROWN CUCKOO-DOVE *M. phasianella* (38 cm) is a red–brown species with mournful, rising and falling 'wow, wow' notes, about four calls in five seconds. The calls of all these cuckoo-doves appear to be very similar, but they can be recognized with practice.

SPOTTED-NECKED DOVE
Streptopelia chinensis (30 cm). Sumatra and Kalimantan.   Figure 3

This turtle-dove is a common inhabitant of open country and even the wooded suburban areas of towns. It is most often seen flying up from the roadsides and bare ground near bushes, when the black-and-white spotted half-collar and broad white tips to the outer tail feathers are clearly visible. The call consists of three or four throaty cooing notes.

3. Spotted-necked Dove

JAMBU FRUIT DOVE
Ptilinopus jambu (27 cm). Sumatra and Kalimantan.   Plate 5d

The Fruit Dove is a quiet but colourful inhabitant of the middle storey of lowland and hill forest. It is usually chanced upon by accident, when a bird makes a brief flight on one's approach, but then perches to give a good view. The female has green under-parts except for a white belly. The white eye-ring is distinctive in both sexes. Generally a silent bird, it will occasionally give a low 'coo', audible only at rather close range.

GREEN-WINGED PIGEON

Chalcophaps indica (25 cm). Sumatra and Kalimantan.   Plate 5e

The Green-winged Pigeon is a common inhabitant of the under-storey, feeding on the ground. It occurs in forest, secondary wood-land, and old rubber plantations, and is often flushed from the side of logging roads. The usual view is of a bird in swift, low flight away from the observer, when the metallic green appearance with the strong white bars on the lower back are diagnostic. It is rarely possible to get a good view of the bird on the ground, but as Plate 5 shows, it is quite colourful, though the female is duller. Hardly a day passes in the forest without its call being heard: a soft call that sounds like 'wup wup wup' uttered quite fast from the under-storey. At very close range, it is heard to consist of a double note, 'cu-wup'.

Parrots

Parrots are characterized by their heavy, hooked bills, used for tearing open fruits and nuts, and also as a grip for climbing along branches. There are very few species in the Sundanese region, however.

LONG-TAILED PARAKEET

Psittacula longicauda (40 cm, with tail). Sumatra and
   Kalimantan.   Plate 6a

Although the male of this parrot is quite colourful, generally it is only seen as a long-tailed green bird in raucous parties in the canopy. It is the commonest member of the family throughout the region, but numbers appear to fluctuate according to poorly understood seasonal movements. Roosting flights over the forest in the evenings may indicate the presence of some quite large roosts, especially near the coasts, such as that at Sukadana on the west coast of Kalimantan.

BLUE-RUMPED PARROT
Psittinus cyanurus (19 cm). Sumatra and Kalimantan. Plate 6b

This short-tailed parrot is not as common as the parakeets, though a few will be heard on most days spent in the forest. The call is a shrill, sharp, whistled note or series of notes, surprisingly musical for a parrot, most often heard from birds in flight at canopy level. The female is a duller bird, with a brownish head, and it lacks the red bill.

BLUE-CROWNED HANGING PARROT
Loriculus galgulus (14 cm). Sumatra and Kalimantan. Plate 6c

The hanging parrots are charming birds, stumpy, almost headless, little bundles of bright green that always rest hanging upside-down. They clamber about the outer branches with great dexterity, but are most often seen in swift flight, giving high, thin 'zri-ie' calls. The female lacks the bright patches of colour of the male, except for the scarlet rump. It is a common species in the lowlands and hills, and not always restricted to forests.

Cuckoos

The cuckoos form a large group, with some 15 species of true cuckoos, the birds which are well known as brood parasites, laying their eggs in the nests of other birds and leaving the raising of their young to the foster parents. In addition, there are 6 species of malkoha and 3 coucals, as well as the very rarely seen Ground Cuckoo. Unobtrusive when not singing, cuckoos are most easily identified from their songs.

ORIENTAL BRUSH CUCKOO
Cacomantis sepulcralis (24 cm). Sumatra and Kalimantan. Plate 6d

Also known as Rufous-breasted or Indonesian Cuckoo, this is a typical member of the group. It is difficult to observe but very vocal. With luck, one can sometimes follow a calling bird and obtain views of it perched on a branch in forest and woodland margins, occasionally in more lightly wooded country and rubber forest. The rather long, graduated tail (rounded when fanned) is typical of the cuckoos, as are the slightly pointed wings. It is dark grey above, paler grey about the foreparts, with a rufous–orange lower breast and belly. However, the Plaintive Cuckoo is very similar. The safest method of identification is from the song. The diagnostic song is a long series of at least ten plaintive high whistles, 'heet, heet, heet, ...', at a constant speed, falling slightly in pitch. However, this song is freely interchanged with another, a three-note phrase, 'tay-ta-wi', repeated about four times up the scale, which is identical to one of the songs of the Plaintive Cuckoo. The Oriental Brush Cuckoo is very common in the lowlands and lower hills of Sumatra, but it appears to be very rare in Kalimantan.

The following notes provide a simple guide to the other cuckoos most often encountered, particularly their calls.

PLAINTIVE CUCKOO
Cacomantis merulinus (22 cm). Sumatra and Kalimantan.

Very similar in colour to the Oriental Brush Cuckoo, the Plaintive Cuckoo has the same rising 'tay-ta-wi' song. In contrast, the diagnostic song begins like that of the Oriental Brush Cuckoo, with three or four 'heet' notes, but then breaks into a rapid cadence of about eight notes down the scale. This is a characteristic song of more open country, including the outskirts of towns, where the bird will sometimes sing in gardens and provide a view on low bushes or electric wires.

BANDED BAY CUCKOO
Cacomantis sonneratii (23 cm). Sumatra and Kalimantan.

The Banded Bay Cuckoo is a bird of the forest canopy, and extremely difficult to see. Its song consists always of four thin, shrill whistles falling slightly but evenly in pitch in a minor key. However, this song should be compared with that of the Indian Cuckoo.

INDIAN CUCKOO
Cuculus micropterus (30 cm). Sumatra and Kalimantan.

The Indian Cuckoo favours the canopy of forest margins, especially in wet places such as river valleys and fresh-water swamps. Like the Banded Bay Cuckoo, it has a song which invariably consists of four notes, but in the Indian Cuckoo's, the notes are much more even and emphatic, three notes almost on one pitch and the fourth, lower. This call is a frequent accompaniment to evenings in a forest camp in the lowlands, and in the wet season it may be heard throughout the night and day.

LARGE HAWK-CUCKOO
Cuculus sparverioides (33 cm). Sumatra.

This bird of forest margins in the mountains is remarkable for its song, from which it has earned the name 'Brain-fever bird', though that name has been applied to several cuckoos. The typical song is a double note, 'pi-pee', that is repeated faster and faster up the scale until it reaches a frantic crescendo. A similar call consists of a single note that is repeated at constant speed rapidly up to a crescendo and then fading off. The two songs can be interchanged at will. It is said that two birds in close proximity singing in competition throughout a moonlit night can drive a person mad. It is found in the mountains of North and West Sumatra, and also in northern Borneo.

VIOLET CUCKOO
Chrysococcyx xanthorhynchus (16 cm). Sumatra and
Kalimantan.   Plate 6e

This common bird of lowland forests is most often seen as a dark cuckoo in song-flight above the forest canopy or across logging roads; the flight is dipping, the dips synchronized with a rather shrill 'kie-vik', one of the characteristic sounds of the forest although easily overlooked by anyone not yet familiar with it.

DRONGO-CUCKOO
Surniculus lugubris (23 cm). Sumatra and Kalimantan.

One of the characteristic calls of lowland forests and woodlands is the human-like whistle of five to seven notes rising up the scale, given by the Drongo-Cuckoo. It is a black, drongo-like bird, with a very slight fork in its rather long tail, which is barred with white.

RAFFLES' MALKOHA
Phaenicophaeus chlorophaeus (33 cm). Sumatra and
Kalimantan.   Plate 7a

The malkohas are not parasitic and appear to have little in common with the true cuckoos. All are characterized by their very long tails. They are sluggish birds of the middle storey that tend to creep around the foliage rather like squirrels, occasionally making short, laboured flights to another bush, on broad, rounded wings.

With six species (only five in Kalimantan), it is difficult to select any one that is typical. The Raffles' Malkoha is one of the commoner species in the lowlands, in both primary and secondary forest. It is the smallest, and the only species in which both the bill and the eye-shield are green. Plate 7 shows the female, whose grey foreparts are very distinctive. The male is mostly chestnut. Most malkohas are generally silent, but this species habitually gives a series of about five notes down the scale, uttered as if with great effort.

Two other species are mostly chestnut. The CHESTNUT-BREASTED *P. curvirostris* (46 cm) has a green bill and red eye-shield, chestnut breast and belly, a greenish mantle, and no white in the tail. It is common in open scrub. The RED-BILLED MALKOHA *P. javanicus* (46 cm) is the only species with a red bill, and it has no obvious eye-shield.

CHESTNUT-BELLIED MALKOHA
Phaenicophaeus sumatranus (40 cm). Sumatra and
Kalimantan.   Plate 7b

The other three species are mainly grey, with green bill, red eye-shield, and white tips to the underside of the tail feathers. The grey is quite dark, with a greenish gloss. The chestnut under tail coverts that identify this species are lacking in the very similar BLACK-BELLIED MALKOHA *P. diardi* (37 cm), while the GREEN-BILLED MAL-KOHA *P. tristis* (56 cm) is a much larger bird of the hills that is present in Sumatra but not Kalimantan.

GREATER COUCAL
Centropus sinensis (53 cm). Sumatra and Kalimantan. Plate 7c

The coucals or crow-pheasants are rather heavy, clumsy birds with medium-length tails. The Greater Coucal is common in dense thickets, whether in primary or secondary forest or damp places such as the edge of ricefields. It is readily identified, with its thick bill, black-and-chestnut plumage, and graduated black tail. It is a skulking bird, often feeding on the ground, and its local name of 'bubut' is derived from the call, a number of hollow 'hoop' notes. The full song consists of a series of about 30 such notes, beginning loud and quite fast, but soon tailing off into a long series of even notes.

In thin secondary scrub and bushes in dry grassland, one finds the LESSER COUCAL *C. bengalensis* (38 cm), a rather more scruffy bird with dark brown or black body, much spotted and streaked with white. The usual call is three or four 'hoop' notes that break into a staccato call, quite distinct from that of the Greater Coucal.

Owls

The mysterious owls and other night birds are intriguing, but difficult to study. Owls range in size from the small scops owls to the large eagle-owls. All of them are primarily arboreal and nocturnal, but an observer will sometimes chance upon an owl during the daytime, usually sitting silently, suffering with as much resignation as it can the mobbing of small birds that often reveals its presence. One of the disconcerting features of owls is their ability to rotate their heads, always to face the intruder, with their big round eyes set in a flat facial disk.

1. (a) Storm's Stork, p. 3. (b) Milky Stork, p. 2. (c) White-
shouldered Ibis, p. 3. (d) Cotton Pygmy Goose, p. 4.

2. (a) Lesser Fish-Eagle, p. 7. (b) Black-shouldered Kite, p. 6.
   (c) Brahminy Kite, p. 6. (d) Crested Serpent-Eagle, p. 7.
   (e) Black-thighed Falconet, p. 8.

3. (a) Great Argus, p. 8. (b) Crested Fireback, p. 8. (c) Blue-
breasted Quail, p. 9.

4. (a) White-breasted Waterhen, p. 9. (b) Slaty-breasted Rail, p. 10.
(c) Pheasant-tailed Jacana, p. 10. (d) Black-naped Tern, p. 12.

5. (a) Little Green Pigeon, p. 12.   (b) Cinnamon-headed Pigeon, p. 13.
   (c) Little Cuckoo-Dove, p. 13.   (d) Jambu Fruit Dove, p. 14.
   (e) Green-winged Pigeon, p. 15.   (f) Green Imperial Pigeon, p. 13.

6. (a) Long-tailed Parakeet, p. 15.   (b) Blue-rumped Parrot, p. 16.
(c) Blue-crowned Hanging Parrot, p. 16.   (d) Oriental Brush
Cuckoo, p. 16.   (e) Violet Cuckoo, p. 18.

7. (a) Raffles' Malkoha, p. 19.   (b) Chestnut-bellied Malkoha, p. 19.
   (c) Greater Coucal, p. 20.   (d) Whiskered Treeswift, p. 25.

8. (a) Buffy Fish-Owl, p. 21. (b) Brown Hawk-Owl, p. 21.
(c) Collared Scops Owl, p. 21. (d) Malaysian Eared Nightjar,
p. 22. (e) Javan Frogmouth, p. 22.

COLLARED SCOPS OWL

Otus bakkamoena (22 cm). Sumatra and Kalimantan.   Plate 8c

The Collared Scops Owl is the only species that occurs commonly in open country, and its single, slightly querying, 'wok' calls, uttered at irregular intervals, are often heard in wooded villages or the tree-lined suburbs of towns. A torchlight will reveal the yellowish-brown eyes and short ear tufts.

BROWN HAWK-OWL

Ninox scutulata (30 cm). Sumatra and Kalimantan.   Plate 8b

Perhaps the most frequently heard owl in forest is the Hawk-Owl, so called because it is rather hawk-like in appearance, lacking the flat facial disc of most owls. It has a hawk's powerful hooked bill and glaring yellow eyes. It lives in all kinds of forest in the lowlands and lower hills, where its call can be heard most nights: a repeated double note, 'wu-wup, wu-wup, wu-wup', the second note higher and emphasized, uttered more rapidly than the irregular, nearly mono-syllabic calls of the Collared Scops Owl.

BUFFY FISH-OWL

Ketupa ketupu (45 cm). Sumatra and Kalimantan.   Plate 8a

Among the larger owls, perhaps this is most often seen, as it favours the more open wooded localities near water, sometimes flying along-side rivers or swamps, and even occasionally ricefields in wooded country. It feeds on fish, frogs, and other water creatures, and its legs are unfeathered, appropriate to a watery environment. It is a rich rufous colour with heavy dark vertical streaks. Two prominent ear tufts lie on the side of the head; these are tufts of feathers, not ears.

In the forest, the similar BARRED EAGLE-OWL *Bubo sumatranus* (45 cm) has greyer plumage, prominent ear tufts, and fine horizontal bars on the breast. It is one of the authors of deep owl hoots in the forest.

The MOUNTAIN SCOPS OWL *Otus spilocephalus* (20 cm) may be heard in the mountains of Sumatra, with its plaintive double whistle, 'plew, plew'. This species has not yet been recorded from Kalimantan, but it occurs in neighbouring Sabah and Sarawak.

Nightjars

Nightjars are nocturnal aerial feeders, with long pointed wings, rather long tails, and acrobatic, hawk-like flight. The rich, heavily mottled brown plumage is a good camouflage for birds resting on the ground or along branches, usually lengthwise, during the day. The small bill opens to reveal a wide gape for catching insects in flight.

MALAYSIAN EARED NIGHTJAR
Eurostopodus temminckii (27 cm). Sumatra and
Kalimantan.   Plate 8d

This is the common forest nightjar of the lowlands and lower hills. It lives in the forest, and precisely at dusk and again at dawn, all the eared nightjars in the district will be in flight, above the canopy or around clearings, giving their unmistakable 'tok, tedau' calls. The dusk call comes just as the sunset shrilling of the cicadas dies down, but before full darkness sets in. Unlike other nightjars and owls, they are never heard through the night. Note the prominent ear tufts and pale buff collar in this species.

There are two open-country species. The LARGE-TAILED NIGHTJAR *Caprimulgus macrurus* (30 cm) has a monotonous, emphatic, and resonant 'tonk-tonk, tonk-tonk-tonk' call. It occurs in lightly wooded country, but it is not very common, and appears to be rare in Kalimantan. The SAVANNA NIGHTJAR *C. affinis* (23 cm) occurs widely in open country in Sumatra, and sometimes in towns, though it is confined to the south-east corner in Kalimantan; this has a querying 'schwik' call uttered in flight.

JAVAN FROGMOUTH
Batrachostomus javensis (24 cm). Sumatra and
Kalimantan.   Plate 8e

The frogmouths are an interesting nocturnal family about which very little is known. Superficially similar to nightjars, they derive their name from the bill, which is of almost ridiculous width. Five species occur in the region, and all are a rich, mottled rufous in colour. Often silent birds, the calls are poorly known, but they are a guide to identification. The Javan Frogmouth has a ventriloquial, mournful, soft, falling whistle, wavering slightly at the end, uttered at intervals through the night. Mainly a forest bird, it has also been sighted by the author in secondary growth and among old rubber trees.

Swifts

Swifts are aerial insect-feeders that spend the greater part of their lives in the air. So common at any one place by day, it is often a puzzle where they spend the nights. Being masters of the air, true swifts have very weak legs, and never perch like swallows on wires, branches, or rooftops, but cling to vertical surfaces. The largest swifts are capable of extremely rapid flight and one may be surprised by the 'whoosh' of a passing needletail. Mainly blackish birds, they have curved, sickle-shaped wings, and fly with alternating wing-beats and tilting glides, never closing the wings like the superficially similar swallows.

The small swiftlets and the large needletails are extremely difficult to identify, and even their taxonomy is not clearly understood. It is from some of the small swiftlets that is obtained the highly prized bird's nest soup (nests constructed of saliva), and the cave-dwelling members of the group are capable of echo-location, a sort of avian radar. A good time to watch swifts is when one chances upon them dipping down to the surface of rivers or ponds to drink.

WHITE-BELLIED SWIFTLET
Collocalia esculenta (10 cm). Sumatra and Kalimantan.

The White-bellied is the only swiftlet that is easy to identify, by its white belly and the bluish gloss on the upper-parts. It is common everywhere, but is one of the species that do not build edible nests.

The PALM SWIFT *Cypsiurus balasiensis* (13 cm) is a very slender swift, dark brown with very narrow, curved wings and rather long, forked tail, that is generally seen around palm trees in open country.

HOUSE SWIFT
Apus affinis (15 cm). Sumatra and Kalimantan.   Figure 4

This is the common swift of towns everywhere, with
sometimes hundreds nesting in the eaves of
downtown shop-houses. A blackish bird,
it has slender, curved wings and a slightly
forked tail. The rump and throat are white.
It is a noisy bird, and in the evenings
especially, the streets are full of whirling,
twittering birds. The largest gathering
seen by the author was at Padang   4. House Swift
Sidempuan, but similar numbers are not unusual
in many towns.

SILVER-RUMPED SWIFT
Rhaphidura leucopygialis (12 cm). Sumatra and
    Kalimantan.   Figure 5

The needletails are more heavy-bodied swifts, with
'butter-knife' wings (sometimes appearing to be
broader towards the tip) and short, rounded tails,
the feathers of which terminate in small spines.
The Silver-rumped Swift is the smallest of
the group and is a common resident in
forested areas. It is glossy black with
shining white rump and upper tail
coverts, and it is easy to identify.
Among the larger needletails, the
BROWN NEEDLETAIL *Hirundapus
giganteus* (25 cm) is resident. These are
the swifts whose passage recalls that
of a jet plane: one hears the noise of
the wings before seeing the birds!

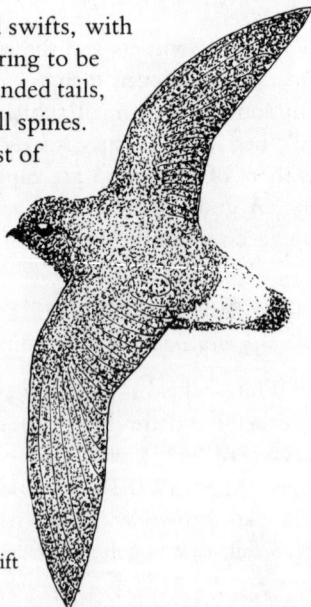

5. Silver-rumped Swift

WHISKERED TREESWIFT

Hemiprocne comata (15 cm). Sumatra and Kalimantan.   Plate 7d

The treeswifts are not true swifts but they have a similar silhouette in flight. Whiskered Treeswifts habitually perch in trees in the forest or on the edge of clearings, and sally forth after insects, returning to perch and folding their wings in jerky movements, as if by numbers, like soldiers on parade closing arms. At rest, the very slender wing-tips cross each other above the long forked tail. Usually in twos or threes, sometimes they gather to roost in quite large numbers at favoured places. As Plate 7 shows, they are handsome and quite colourful birds.

The GREY-RUMPED TREESWIFT *H. longipennis* (20 cm) spends longer in the air, and thus wanders further from the forest. The under-parts are pale, and the very long, needle-forked tail is distinctive. A large crest is visible when the bird is perched. This species also occurs in Java and Bali.

Trogons

The trogons are some of Indonesia's most colourful forest birds, but they tend to sit quite quietly in the middle storey of dense forest, so it is difficult to obtain good views. When disturbed, they will usually fly a short distance to a new perch, with just a tantalizing tip of the tail left to view. The wings are short and rounded, and the tail is moderately long and square-tipped. The females lack the very bright colours which serve to identify the males.

RED-NAPED TROGON

Harpactes kasumba (32 cm). Sumatra and Kalimantan.   Plate 9a

This is a characteristic trogon of dense lowland forest, in which the male has a black head and upper breast, a pronounced red collar on the nape, and a crescentic white band separating the black breast from the scarlet abdomen. It is usually seen in pairs, but more often only its call is heard—a series of about six or seven slow 'kau' notes on one pitch.

The very similar DIARD'S TROGON *H. diardii* occurs in the same habitat. It differs in having a pinkish hind collar, and only a narrow, pinkish breast band, while the top of the crown is maroon. The very similar call differs in consisting of eight to ten notes, faster and slightly falling in pitch.

SCARLET-RUMPED TROGON
Harpactes duvaucelii (24 cm). Sumatra and Kalimantan.   Plate 9b

The Scarlet-rumped Trogon is smaller, and only the head is black, both the breast and belly being scarlet, as is the rump. Being less confined to the interior of dense forest, it is more readily seen than the previous species, and it is also more vocal. The song is a cadence of some dozen notes beginning quite slowly and accelerating rapidly.

The common species in the mountains of Sumatra is the BLUE-TAILED TROGON *H. reinwardtii* (30 cm), which is mainly green and blue-green, with a yellow throat and belly, and a red bill.

Kingfishers

Kingfishers are colourful birds but often shy and difficult to see, except as a dash of colour vanishing around the next bend in the river. There are at least a dozen species, and they occur in a range of habitats, not always near water, in both forest and open country. Those living within the forest are especially difficult to see, as they will sit quietly through much of the day, in contrast to some of their noisy open-country cousins. The more commonly seen kingfishers are treated below, from the smallest upwards.

ORIENTAL DWARF KINGFISHER
Ceyx erithacus (14 cm). Sumatra and Kalimantan.   Plate 9f

These tiny kingfishers can often be seen by small rivers and in swampy areas in the forest, usually as a little darting reddish ball with a long beak. There are two forms, often treated as separate species, though hybrids occur. The redder form is more common, and is a lovely lilac mauve colour, brighter on the rump, and the under-parts are yellow. The darker form has a blackish mantle and wings; this one is also migratory and so may turn up in a variety of habitats. The call is a soft, high squeak.

COMMON KINGFISHER
Alcedo atthis (18 cm). Sumatra and Kalimantan.   Plate 9e

The Common Kingfisher of Europe occurs widely in winter through-out the region, in open swamps, on river banks, and along coasts. The rufous ear coverts distinguish it from the resident DEEP BLUE KINGFISHER *A. meninting* (15 cm), which is much deeper blue above, has blue ear coverts, and is found along rivers in forested country.

WHITE-THROATED KINGFISHER
Halcyon smyrnensis (28 cm). Sumatra.   Plate 9d

This is the big kingfisher that occurs in open country throughout Sumatra. A recent arrival from mainland Asia, it was first recorded in Sumatra in 1921, but has since spread throughout that island. The chestnut, white, and blue plumage, with a red bill, is distinctive, and there is a white patch visible on the wings in flight. The call is a loud and raucous slow descending trill. It is by no means confined to water and is not normally a fish-eater.

The WHITE-COLLARED KINGFISHER *H. chloris* (24 cm) is also widespread. Originally a coastal bird, it has spread inland, though it requires wooded country such as rubber estates and wooded villages. It has greenish-blue upper-parts with a white collar, and white under-parts. There is also a rather beautiful species that migrates from northern Asia from September to March, the BLACK-CAPPED KINGFISHER *H. pileata* (31 cm). Seen mainly along the larger rivers, it has a prominent black cap, white collar, and whitish under-parts that become rufous on the belly. However, when viewed in flight from the rear, it is very similar to the White-throated Kingfisher.

STORK-BILLED KINGFISHER
Pelargopsis capensis (37 cm). Sumatra and Kalimantan.   Plate 9c

The largest kingfisher is unmistakable, with its huge red bill, blue wings (without a white patch), and bright blue back. The head is brown and the under-parts buffy. Mainly a coastal bird, it also extends far up rivers and into open fresh-water swamps, and in Kalimantan especially, it is seen along quite small rivers well into the forest. The calls are very loud and rather wailing, though not always unmusical.

Bee-eaters

Bee-eaters are so named because some species include flying bees as a regular part of their diet, but they will take any flying insects. They are slender, graceful birds, and, like nightjars, swifts, and swallows, they are skilful aerial hunters. The bill is fine and slightly curved, and the wings have a rather triangular shape.

BLUE-THROATED BEE-EATER
Merops viridis (28 cm). Sumatra and Kalimantan.   Plate 10a

The Blue-throated Bee-eater is most often seen along logging tracks and rivers in thickly wooded terrain and forests, hawking for insects at or below tree-top level in twos or threes, coming to rest periodically on exposed branches, but never really sitting quietly. The call is a soft 'kwilp, kwilp'. It nests in burrows in banks, singly or in groups, and availability of sandy banks defines where it is found in the breeding season. Thus coastal and riverine sites are especially favoured. Outside the breeding season, it is often seen in small flocks in more open country, and it migrates locally.

   Between August and April, it can be confused with the BROWN-THROATED BEE-EATER *Merops philippinus* (30 cm), which is a common migrant. This differs in having a green head, and a pale yellow throat bordered below by a brown gorget on the breast. Both species have a pale blue rump and tail.

RED-BEARDED BEE-EATER
Nyctyornis amictus (33 cm). Sumatra and Kalimantan.   Plate 10b

The Red-bearded Bee-eater is a much more sluggish bird that never leaves the forest. Plate 10 shows the prominent red throat and breast, and lilac forehead. The harsh call of four or five descending 'ka-ka-

ka-ka' notes readily reveals its presence, the birds usually perched in twos or threes in the middle storey.

Rollers

DOLLARBIRD
Eurystomus orientalis (30 cm). Sumatra and Kalimantan.   Plate 10c

Rollers are so named from their rolling flight in courtship display. The Dollarbird (also called Broad-billed Roller) is a rather solid dark bird, with red legs and thick, red bill, and acquires its name from a circular, pale silvery blue patch on the wings, visible in flight. It is seen in wooded areas and forest clearings, singly or in pairs on exposed perches, from which it watches for insects in the air, or occasionally on the ground. The voice is harsh and rasping.

Hornbills

Even people with no interest in birds will enquire about the hornbills, and indeed they are very impressive, with their massive build, prominent bills, and loud calls. There are no less than nine species (eight in Kalimantan). They are broad-winged and long-tailed, and the larger species make quite long flights from one patch of forest to another in their search for trees in fruit. The wings of some species are very noisy, like the 'whoosh, whoosh' of a steam engine, often the first indication that hornbills are flying overhead. All the hornbills have a distinctively shaped horny protuberance or casque above the bill. They nest in holes in the trunks of trees, and seal up most of the entrance hole with mud, thus imprisoning the female during incubation, presumably as a protection against monkeys.

RHINOCEROS HORNBILL
Buceros rhinoceros (male 120 cm, female 90 cm).
   Sumatra and Kalimantan.   Plate 10d

This is perhaps the best known of the hornbills. It is a large, noisy bird, black, but with a white belly, and a white tail with a black bar across the middle. The up-turned casque is yellow in front, reddish behind. The call is a series of single loud honks, 'hok, hok', which usually break into a raucous, goose-like 'g-raak, g-raak' as the bird takes to flight, audible from afar.

PIED HORNBILL
Anthracoceros coronatus (75 cm). Sumatra and Kalimantan.   Figure 6

The smallest hornbill is the one most seen by those who do not customarily enter the forests, for it occurs in wooded areas and forest margins along rivers, swamps, and coasts. Figure 6 shows the black-and-white plumage. The rear edge of the wings and the outer tail feathers are white, conspicuous in flight. Although their calls are not as loud as those of the previous species, they are shrill and vociferous birds with a variety of yelping calls: 'yip, yip, yip, ayip, youk, youk', and so on.

The BLACK HORNBILL *Anthracoceros malayanus* (75 cm) is black, except for the white tips to the outer tail feathers, and the bill and casque which are white in the male. The call is a coarse, rasping note that recalls the squealing of an angry pig.

6. Pied Hornbill

WRINKLED HORNBILL
Rhyticeros corrugatus (80 cm). Sumatra and Kalimantan.   Plate 10e

The Wrinkled Hornbill is black, but with white head and neck, and black crown and small crest. The basal third of the tail is black, and the rest white, though the white parts are often stained dirty yellowish-brown. There is a white gular pouch below the bill.

The WREATHED HORNBILL *Rhyticeros undulatus* (100 cm) is very similar, and the larger size is not a useful feature in field identification. However, the whole of the tail is white (though again, usually stained), the crown and crest are chestnut, and there is a clearly visible black bar across the gular pouch, which is yellow in the male but pale or bluish in the female. When seen in overhead flight, the two species are best identified by the tail and gular pouch.

The calls of both the Wrinkled and Wreathed Hornbills are harsh barking notes, but they are not very vocal. Both have very noisy wings in flight.

HELMETED HORNBILL
Rhinoplax vigil (125 cm, female smaller, but both have at least
an additional 20 cm of elongated central tail feathers).
Sumatra and Kalimantan. Plate 10f

This stunning bird is a favourite to anyone who has heard its call. It is huge and unmistakable, particularly with its long tail feathers. The extraordinary call begins with slow, single 'hoop' notes at long intervals, eventually speeding into disyllabic notes—'te-hoop'—which build to a crescendo that then breaks into a cadence of 'poo' notes, abruptly terminated by a rapid, descending, cackling laugh. The whole (abbreviated) sequence, which may take several minutes, is as follows: 'hoop—/—hoop—', etc., 'te-hoop, te-hoop, te-hoop, poo, poo, poo, poo, ka-ka-ka-ka-ka'. Commonly, the second bird will both begin and end its sequence shortly behind that of its mate.

This hornbill is confined to the hills in Kalimantan, but in Sumatra it also extends widely across the lowlands.

Barbets

The main feature of many barbets is the relentless persistence of their calls, so much so that after a while, an observer no longer notices them. The heads are turned whilst calling, and the bills remain closed, giving a ventriloquial effect of varying volume.

Barbets are stocky birds with heavy bills and short tails, and they are poor fliers, flying only short distances on audibly whirring wings. Most are predominantly green but with a startling array of gaudy colours in the head region, which are diagnostic when visible, but this is unusual as the birds are generally hidden in the canopy.

Six species occur in both Sumatra and Kalimantan. In addition, Sumatra has three species that also occur in the Malay Peninsula, while the island of Borneo has three montane endemics.

Barbets are recognized mainly from their calls, although some species have two or even three quite different calls, which is confusing.

RED-CROWNED BARBET
Megalaima rafflesii (25 cm). Sumatra and Kalimantan.   Plate 11a

The Red-crowned Barbet is one of the characteristic barbets of the lowland forests. The call consists of two notes, 'tuk, tuk', a pause, and then some 10 to 20 such notes delivered at constant speed for a few seconds. The author vividly recalls fine evenings in the Jambi forest when the only sounds were the calls of this bird from all directions, mellow in the still air. The pitch and speed of delivery may differ between individuals, but the pause after the second note is absolutely unvarying.

Plate 11 illustrates the Red-crowned Barbet in full, but only the head patterns of the other barbets are shown. The calls and habitats are described below.

BLUE-EARED BARBET
Megalaima australis (18 cm). Sumatra and Kalimantan.   Plate 11e

Probably the commonest barbet in the lowlands and lower hills, the Blue-eared has also the most monotonous call, seemingly repeated for hours at a time. There are three types of calls, the most typical being the endless disyllabic 'tr-trrk' repeated at about two per second. Another call is monosyllabic, uttered at the same speed or slightly faster, but only for short periods, which has been likened to the sound of a football referee's whistle. A third call is a rather tinny rolling note, repeated at shorter intervals and quickening in speed.

GOLD-WHISKERED BARBET
Megalaima chrysopogon (30 cm). Sumatra and
  Kalimantan.   Plate 11c

The call of this species is a loud 'te-hoop', repeated rapidly for many
minutes at a time. There is also a trilling call, beginning with a long
trill that is repeated in gradually shortening phrases, until they break
into a long series of triple and finally double notes. In Kalimantan this
barbet is confined to the more hilly country, but in Sumatra it
extends quite far into the plains.

RED-THROATED BARBET
Megalaima mystacophanos (23 cm). Sumatra and
  Kalimantan.   Plate 11b

Another species of the lowlands and lower hills, but this is not quite
so abundant. The call of this species consists of 'tuk' notes uttered in
irregular phrases, typically one, two, or three at a time, with short
breaks. It also has a rolling or trilled note, repeated in shortening
phrases, similar to that of the previous species.

YELLOW-CROWNED BARBET
Megalaima henricii (22 cm). Sumatra and Kalimantan.   Plate 11d

This lowland barbet has an unmistakable call that consists of a roll
and four 'tuk' notes, repeated: 'trrrrk-tuk-tuk-tuk-tuk, trrrrk-tuk-
tuk-tuk-tuk', etc.

COPPERSMITH BARBET
Megalaima haemacephala (15 cm). Sumatra.   Plate 12a

The smallest barbet, this has a regular 'tonk, tonk' call from which it
gets its name. This is not a forest bird, being found in wooded areas
around traditional Sumatran villages, but it does not seem to be very
common. Birds in Java are a different sub-species, having a red throat
and breast, and lacking the yellow.

BLACK-BROWED BARBET
Megalaima oorti (20 cm). Sumatra.   Plate 12b

The Black-browed Barbet is a montane species, heard commonly throughout the higher mountains at an elevation of 1 000 to 2 000 m. The call is a repeated, rapid three-note phrase, the second note softer and the third rolled: 'took-tk-trrrk', repeated about once per second.

FIRE-TUFTED BARBET
Psilopogon pyrolophus (28 cm). Sumatra.   Plate 12c

Another montane barbet, the Fire-tufted has an extraordinary cicada-like buzzing note, quite unlike the repetitive 'took's and 'poop's of the other species. The buzz begins as single notes which gather speed until it ends in a longer, harsh buzz, the whole phrase lasting about three seconds.

MOUNTAIN BARBET
Megalaima monticola (23 cm). Kalimantan.   Plate 12d

Few bird-watchers have the good fortune to reach the higher mountains of Borneo on the Indonesian side of the border, where interesting discoveries await them. For example, the Mountain Barbet has been reported only recently on Gunung Niut in West Kalimantan. The call of this barbet is a rapid series of 'took' notes, usually very fast, but with frequent and irregular short pauses.

Woodpeckers

Woodpeckers form a big family, with up to 23 species in Sumatra, of which 16 occur in Kalimantan. They are adapted to feed on insects and wood-borers hidden in crevices in the bark of trees, for which they have strong, pointed bills. Their stiff tails are used as a support while climbing vertical tree trunks. They nest in holes which they have bored or enlarged themselves. When feeding, the tapping of bill on wood will often attract attention, and some species 'drum' very rapidly with their bills as a form of territorial vocalization. They are quite strong fliers, with characteristic dipping flight on short, rounded, whirring wings. All are resident, and many species range widely in forests and woodland edges through the lowlands and hills. Most have sharp, chittering, or shrill call notes, which are not very helpful for identification.

It is difficult to make a selection from so many species, and only the commoner ones are described, divided into four groups.

1. Small woodpeckers (13–22 cm), having patterned or barred plumage with few bright colours, often found in lightly wooded terrain:

BROWN-CAPPED WOODPECKER
Dendrocopus moluccensis (13 cm). Sumatra and
Kalimantan.   Plate 12e

This small, unobtrusive woodpecker occurs in wooded growth especially near the coasts. Mainly brown and white, the head gives a striped appearance, and the male has a red streak in the cheek (not illustrated in Plate 12). It is seen on low bushes or the lower branches of trees, including village fruit plantations, and its soft but quite forceful chittering notes often reveal its presence.

GREY-AND-BUFF WOODPECKER
Hemicircus concretus (13 cm). Sumatra and Kalimantan.   Plate 12f

This more distinctly marked woodpecker has a pronounced triangular crest. The crown is crimson, the back has a striking scalloped pattern of black and buff, and the rump is white.

Two slightly larger species have strongly barred plumage. The BUFF-RUMPED WOODPECKER *Meiglyptes tristis* (19 cm) is dark with strong whitish bars and rump, and a red cheek patch in the male. It has a triangular crest but this is not often raised. The BUFF-NECKED WOODPECKER *Meiglyptes tukki* (22 cm) is more finely barred, with a pale buff patch on the sides of the neck, but not on the rump. Again, the male has a red cheek patch, but there is no crest.

All three species are quite common in lightly wooded country and forest edges. They have rather passerine behaviour, often perching across the branches, and all have shrill chittering or short trilling calls.

2. Medium woodpeckers (25–34 cm), having green plumage with extensive red and yellow, belonging to the genus *Picus*:

CHECKER-THROATED WOODPECKER
Picus mentalis (28 cm). Sumatra and Kalimantan.   Plate 13a

One of the three common green woodpeckers of lowland forests, the Checker-throated has a yellow crest, and no red on the head. It has a reddish-chestnut collar and wings, and a black-and-white checkered throat.

The CRIMSON-WINGED WOODPECKER *Picus puniceus* (25 cm), is green with a red crest, tipped yellow, and also has red wings. The BANDED WOODPECKER *P. miniaceus* (25 cm) has a similar colour configuration but with an all-red head, and strong barring especially on the under-parts.

All three have similar calls, variously described as 'shay' or 'sheok', either singly or in runs of a few notes. It is difficult to distinguish them by their calls, perhaps because these are so often uttered at dawn or dusk when the bird is silhouetted at the top of a tree and the colours cannot be seen.

LESSER YELLOW-NAPED WOODPECKER
Picus chlorolophus (27 cm). Sumatra.   Plate 13c

The Lesser Yellow-naped Woodpecker replaces the previous three species in the mountains of Sumatra. The yellow crest is particularly striking in this species, but the GREATER YELLOW-NAPED WOODPECKER *P. flavinucha* (34 cm), which occurs in the same habitat, is rather similar. However, the latter bird lacks any red on the head, and has banded black and reddish wings and unmarked green under-parts.

3. Medium woodpeckers (23–25 cm) without green in the plumage:

MAROON WOODPECKER
Blythipicus rubiginosus (23 cm). Sumatra and
Kalimantan.   Plate 13b

The unbarred maroon upper-parts and pale yellow bill are distinctive on this bird. However, the RUFOUS WOODPECKER *Celeus brachyurus* (25 cm) is quite similar, being dark rufous with some black barring, but with a dark bill; the male has a red cheek patch. Both are quite common in a range of forested and wooded habitats.

4. Larger woodpeckers:

COMMON GOLDEN-BACKED WOODPECKER
Dinopium javanense (30 cm). Sumatra and Kalimantan.   Plate 13d

The Golden-backed is a large, prominently coloured woodpecker of secondary growth and gardens, including rubber and coconut plantations. With its golden yellow wings and dipping flight, it might be mistaken for an oriole, but it is more elongated, in addition to having a red crest and rump, banded black-and-white cheeks, and white under-parts with black crescentic markings.

The ORANGE-BACKED WOODPECKER *Reindwardtipicus validus* (30 cm) is a handsome bird, characterized by a long buff stripe from the back of the neck to the rump, giving an elongated, long-necked appearance. The under-parts and crest are red, and there are chestnut bars in the black wing. It has a typical woodpecker chittering call, and one of the author's favourite memories is of a family of these, all drumming in turns in a bamboo grove. Each bamboo resonated at a different pitch, and it is difficult to believe the birds were not as delighted by the sound effects as the observer!

WHITE-BELLIED WOODPECKER
Dryocopus javensis (43 cm). Sumatra and Kalimantan.   Plate 13e

This is the common large woodpecker of the lowland forests, a great black bird with a red crest and white belly. Its call is a very loud, metallic 'sheok' that rings through the forest.

In Kalimantan, there is an even larger woodpecker, the GREAT SLATY *Muelleripicus pulverulentus* (51 cm), easily identified by its large, long-necked appearance and slaty colour, with a buff throat. The loud whickering or whinnying call, 'wooik, wooik' and 'wi-wi-wik', can usually be heard on most days spent in lowland forest, often followed by a view of three or four in flight at canopy height, with large, rounded, dark wings.

Broadbills

The broadbills are a colourful family of arboreal birds, with rather squat bodies, and bills that seem almost ridiculously broad if viewed head-on. Most broadbills are heard far more often than seen, and one or two species will be heard on any day spent in the forest. Once the songs have been learnt, they can be recognized immediately. There are seven species in Sumatra, of which six occur in Kalimantan; in addition, Kalimantan has two endemic species. In contrast, only one species occurs in Java.

BLACK-AND-YELLOW BROADBILL
Eurylaimus ochromalus (15 cm). Sumatra and
Kalimantan.   Plate 14b

The Black-and-Yellow is the broadbill that is most commonly heard in the forests of the lowlands and lower hills. The loud song begins slowly and gathers speed until it reaches a cicada-like buzz, ending abruptly. The plumage, which is the same in both sexes, is very striking, but it is extremely difficult to obtain a good view. Both sexes have the same song, usually one of a pair beginning its song slightly behind its mate, or uttering plaintive 'piu' calls in accompaniment.

BANDED BROADBILL
Eurylaimus javanicus (23 cm). Sumatra and Kalimantan.   Plate 14d

The Banded Broadbill's song is very similar to that of the previous species, except that it begins with a short flourish, followed instantly by the cicada-like series of rapid notes uttered at constant speed, until the final notes which tail off at the end. The female lacks the dark breast band shown in Plate 14. The bill is all blue.

BLACK-AND-RED BROADBILL

Cymbirhynchus macrorhynchos (25 cm). Sumatra and
    Kalimantan.   Plate 14a

This species occurs mainly in riverine forests, and is often seen as it flies across the rivers, particularly in Kalimantan. The plumage is a deeper red than that of the Banded, and it has white instead of yellow in the wing; the bill is blue above and yellow below. It is a much more silent bird, occasionally giving a quiet but quite harsh note.

DUSKY BROADBILL

Corydon sumatranus (28 cm). Sumatra and Kalimantan.   Plate 14e

The largest of the lowland broadbills, this species is dark in colour, with a pale buff throat and a pinkish bill. The song consists of the repetition of a rather lisping 'tiu-wit', four or five times, on a slightly ascending scale.

GREEN BROADBILL

Calyptomena viridis (19 cm). Sumatra and Kalimantan.   Plate 14c

Being all green, this is not an easy bird to see in the foliage of the canopy. As Plate 14 shows, it is a curiously shaped bird, very chunky with the feathers of the forehead partly covering the bill. The female lacks the black in the neck and wings. Its voice is commonly heard in the forests of the lowlands and lower hills, and can be recognized instantly once known: a short bubbled note, emphasized at the start and with an upwards inflexion, having the timbre of a stone bouncing on a frozen lake. There are other calls but this is the most typical.

The LONG-TAILED BROADBILL *Psarisomus dalhousiae* (28 cm) occurs in the mountains of Sumatra. It is mainly green, but this is offset by brilliant colours—yellow about the face and throat, and blue in the wings and on the long tail. The crown is black. The song is a typical broadbill run of notes, beginning suddenly and gathering speed. It also occurs in the mountains of northern Borneo, and should be looked for along the borders of Kalimantan.

Borneo also has two endemic montane broadbills, both of which extend across the border into the mountains of Kalimantan. HOSE'S BROADBILL *Calyptomena hosei* (20 cm) is very similar to the Green Broadbill, but has pale blue under-parts. WHITEHEAD'S BROAD-BILL *Calyptomena whiteheadi* is like a large version of the Green Broadbill but with a black throat.

Pittas

This is an enigmatic family. Although often having brilliant colours, they are secretive birds of the forest floor, and the casual bird-watcher is very lucky if he chances upon one. They are stocky birds, with very short tails, short and rounded wings, and rather long legs. The calls are distinctive, and the best way to see a pitta is to identify the call, and then to stealthily follow it, but this requires time and patience. Despite their secretive behaviour, some species are migratory.

GARNET PITTA
Pitta granatina (18 cm). Sumatra and Kalimantan.   Plate 15a

Although the plumage shown in Plate 15 is distinctive, the Garnet Pitta is a very elusive bird. However, its call, consisting of a soft monotone whistle of slightly over a second's duration, will often reveal its presence. Unfortunately, two other ground-frequenting birds in the forest, the Blue-banded Pitta and the Rail Babbler (see p. 58), have very similar calls, distinguishable only to the practised ear. The author once spent a full hour stealthily following this pitta and imitating its whistle, finally to be rewarded with a brief glimpse as it momentarily hopped into view on a log, a view that was at once sufficient for identification. It is mainly a lowland bird, and in the mountains of western Sumatra, as well as in Sabah, it is replaced by a black-crowned form which is separated as a distinct species, the BLACK-AND-SCARLET PITTA *Pitta venusta*.

The BANDED PITTA *Pitta guajana* (23 cm), which occurs in lowland forest in both Sumatra and Kalimantan, has a short 'pow' note and a low 'crrrrr'. It has broad black and orange–yellow stripes on the head, white on the wings, and a finely banded blackish and orange abdomen. The tail is blue.

LESSER BLUE-WINGED PITTA
Pitta moluccensis (20 cm). Sumatra and Kalimantan.   Plate 15b

Plate 15 shows another colourful bird, but it is also elusive. In flight away from the observer, it appears as a mainly blue bird, with a large white patch on the wings. The call is very distinctive, a double whistle best rendered as 'hoyuk-oyuk'. There are both resident and migrant forms, but it is mainly a migrant to this region.

SCHNEIDER'S PITTA
Pitta schneideri (22 cm). Sumatra.   Frontispiece

Schneider's Pitta is a very enigmatic bird, endemic to the mountains of Sumatra. Up to 1918, it was reported as common, but there was no further record until it was rediscovered on Mount Kerinci in 1988. It should be looked for in montane forest at an elevation of between 1 000 and 2 000 m. The call is reported to be a low, soft, double whistle, rising on the first note and falling on the second.

BLUE-HEADED PITTA
Pitta baudi (18 cm). Kalimantan.   Plate 15c

The Blue-headed Pitta is endemic to the lowland forests of Borneo, but is often overlooked unless its loud 'kiaow' call is known. The female differs in being dark rufous above, paler below, and only the tail is blue.

Another Borneo endemic is the BLUE-BANDED PITTA *Pitta arquata* (15 cm) which is reddish, with a green back and a narrow blue breast band. It occurs in hill forest up to about 1 500 m. Its call is a short, quavering whistle reminiscent of that of the Garnet Pitta.

Swallows

The swallows are well known to most people, being familiar birds of towns and villages. Swallows are superficially like swifts, but the wings are not so long and crescentic, and the flight is different. They are largely aerial, hawking for insects with characteristic swooping flight. They roost commonly along electricity and telegraph wires in downtown areas, and some roosts number many thousands. There are two common species, both more or less blue above and whitish below.

PACIFIC SWALLOW
Hirundo tahitica (14 cm). Sumatra and Kalimantan.　　　　Figure 7

This is the resident swallow, with a pair or two found in every village, usually building a mud nest on the rafters or eaves of houses. It has blue upper-parts, a chestnut throat, and the belly is a rather dirty white. The outer tail feathers are slightly elongated.

The BARN SWALLOW *Hirundo rustica* of Europe and Asia is an abundant winter migrant, greatly outnumbering the resident birds from about late August to early May. The adult has more elongated outer tail feathers, and a cleaner white belly, which is separated from the chestnut throat by a black band. However, many wintering birds are immature, and the two species are then not easy to tell apart.

7. Pacific Swallow

Cuckoo-Shrikes

BAR-WINGED FLYCATCHER-SHRIKE
Hemipus picatus (15 cm). Sumatra and Kalimantan.　　　Figure 8

Flycatcher-Shrikes usually occur in small parties in the
upper canopy, where they are easily overlooked,
though their pleasant but rather indistinctive
call notes will often reveal their presence.
Identification is then not difficult, but
note that another species, the BLACK-
WINGED FLYCATCHER-SHRIKE
Hemipus hirundinaceus, is
identical except that it
lacks the prominent
white wing bars.
The females of
both species have
the black replaced
by brown. Both species occur
widely in lowland and hill forest.

8. Bar-winged Flycatcher-Shrike

PIED TRILLER
Lalage nigra (18 cm). Sumatra and Kalimantan.　　　Figure 9

The other black-and-white member
of the group is a more familiar
bird of the lower and middle
canopy in wooded areas, including
plantations, gardens, and mangroves.
The plumage is quite distinctive: black
above, with a white stripe above the eye
and a grey rump, a white patch on the wing,
and white under-parts. The black is again
replaced by brownish-grey in the female.
The call is a short, slow trilling note on
one pitch, with a timbre not unlike the
chirp of a sparrow.

9. Pied Triller

BLACK-FACED CUCKOO-SHRIKE
Coracina larvata (25 cm). Sumatra.

Figure 10

The true cuckoo-shrikes are larger birds of varying shades of grey. This species is a prominent bird of the higher mountains of Sumatra; it is common also in northern Borneo, and so should be looked for in the mountains of Kalimantan. It is a noisy bird, seen in twos or threes on the edge of woodland.

There are two or three species of cuckoo-shrike in the lowlands, but they tend to be rather elusive, and their identification is based on the shades of grey and the extent of a black mask on the face, so they are not easy birds for the beginner.

10. Black-faced Cuckoo-Shrike

Minivets

Like the cuckoo-shrikes, to which they are related, the minivets are wholly arboreal; they are delightful birds with their long tails and brilliant plumage, usually red in the male and yellow in the female. They characteristically occur in parties of up to a dozen or more birds, which keep together with constant twittering calls, but the parties may be loosely strung, with a tendency to fly from tree to tree in twos or threes.

FIERY MINIVET
Pericrocotus igneus (15 cm). Sumatra and Kalimantan.

Plate 15d

This is the smaller of the two species occurring in lowland forests. The plumage is distinctive, but the males of both this and the larger SCARLET MINIVET *Pericrocotus flammeus* (19 cm) are almost identical. As size is not always a reliable guide, it is necessary to study the females for identification. In both species, the females have the red parts replaced by yellow, but in the Fiery Minivet the rump is orange.

MOUNTAIN MINIVET
Pericrocotus solaris (19 cm). Sumatra.   Plate 15e

Further north in Asia, this species has a grey chin, but in the Sundanese region, this can be almost black, and the bird is very similar to the Scarlet Minivet. Altitude is the best guide, though there may be some overlap at around 1 000 m. This minivet is common in the mountains of Sumatra, and also in northern Borneo though it has not yet been recorded from Kalimantan.

In common with Java, the mountains of Sumatra also have the SUNDA MINIVET *Pericrocotus miniatus*, which is again very similar, though in this species the females are reddish, not yellow. It frequently occurs in quite large flocks.

Ioras and Leafbirds

The ioras and leafbirds are mainly green birds that live in the foliage of the canopy, in woodland and forest.

COMMON IORA
Aegithina tiphia (15 cm). Sumatra and Kalimantan.   Plate 15f

This small bird is common in a variety of lightly wooded habitats, including mangroves, and is best known by its call, typically 'weeeeeeee-tu' and variations. A similar bird is the GREEN IORA *A. viridissima* (14 cm), which is a much darker green, especially on the under-parts, while the wings are more black. However, the Green Iora tends to keep to the forest canopy and is more difficult to observe.

GREATER GREEN LEAFBIRD
Chloropsis sonnerati (20 cm). Sumatra and Kalimantan.   Plate 15g

As the name implies, the leafbirds are readily camouflaged in the canopy and not easy to see. There are five species in Sumatra, but only three in Kalimantan. This is the largest of the group, and it has quite a loud and musical song. The black throat of the male is lacking in the female. However, the two other lowland species are very similar, and their identification is not easy.

The BLUE-MASKED LEAFBIRD *C. venusta* (14 cm) is endemic to the higher hills of Sumatra. In the male, the forehead, sides of the head, and tail are blue, and the black throat passes to golden yellow on the upper breast. In the female, the throat is also blue.

Bulbuls

The bulbuls form one of the principal families of smaller birds, with no less than 27 species in Sumatra and 23 in Kalimantan. They are quite noisy birds, usually found in the middle storey, and are characterized by soft and rather fluffy plumage. While most are forest birds, a few are found only in open country. Some bulbuls have distinctive plumage, but many are very difficult to identify. Only a representative selection is described here.

YELLOW-VENTED BULBUL
Pycnonotus goiavier (20 cm). Sumatra and Kalimantan. Plate 16a

This common bulbul of the open country is a very unprepossessing species, with its dowdy plumage and rather thin though cheerful song. It is easily identified by the white stripes above and below the eye, and yellow under tail coverts.

The common species of Java, the SOOTY-HEADED BULBUL *P. aurigaster*, has been introduced into Sumatra, and now occurs quite widely there; it has a black head, and a more golden yellow under the tail. The ORANGE-SPOTTED BULBUL *P. bimaculatus* of Java also occurs in the mountains of Sumatra; this has an orange spot in front of the eye and yellow ear coverts, and again there is yellow under the tail.

STRAW-HEADED BULBUL
Pycnonotus zeylanicus (29 cm). Sumatra and Kalimantan. Plate 16b

The Straw-headed Bulbul is a popular cage-bird, renowned for its very loud and musical song, though the song phrases lack variety. It is a large and distinctive bulbul of damp thickets in wet areas. Probably most of the caged birds sold in Jakarta come from outside

Java, where it is now very rare, so this trade needs to be carefully monitored, or the wild Indonesian population could quickly become endangered.

BLACK-HEADED BULBUL

Pycnonotus atriceps (18 cm). Sumatra and Kalimantan.   Plate 16c

Among the forest bulbuls, the neatly plumaged Black-headed is common in both islands. The main features of this bird are the yellow tip to the black tail, and the yellow lower back. It has a pleasant, musical song, though the phrases do not vary.

RUBY-THROATED BULBUL

Pycnonotus dispar (19 cm). Sumatra.   Plate 16d

This lovely bulbul occurs in the same habitat as the previous species, and differs in having a crest, and a ruby throat; it lacks the yellow in the wings and tail of the previous bird. In Kalimantan, it is replaced by the BLACK-CRESTED BULBUL *P. melanicterus*, which is identical except that the ruby patch is lacking, the entire head including the throat being black.

RED-EYED BROWN BULBUL

Pycnonotus brunneus (18 cm). Sumatra and Kalimantan.   Plate 16e

The Red-eyed Brown is representative of the group of brown bulbuls, common in the forests, which are singularly lacking in distinctive features. At least two other species are almost identical, and even the colour of the eye-ring, from which this species gets its name, is not a reliable feature; identification is best left to the experts. The brown bulbuls have short but quite sweet and musical songs.

YELLOW-BELLIED BULBUL

Criniger phaeocephalus (20 cm). Sumatra and Kalimantan.   Plate 16f

The brown bulbuls are not the only troublesome species, as there is also a group of more olive-coloured bulbuls, including those in the genus *Criniger*. These are usually slightly larger, and harsh of voice.

The Yellow-bellied is one of the easier species to identify, with its clean white throat and yellow under-parts. The head is grey and lacks a crest. Two similar species have rather duller plumage, but both sport a short crest, and both have notably harsh songs. They all occupy the same habitats, forests of the lowlands and lower hills. The *Criniger* bulbuls will often puff out their white throat feathers in display.

Drongos

These are black birds with rather long and strikingly forked tails. They are notoriously aggressive, driving off birds twice their size, even eagles and hornbills. They have a wide range of call notes, sometimes musical, and occasionally mimicking other species, and the calls are not much help in identification.

BRONZED DRONGO
Dicrurus aeneus (24 cm). Sumatra and Kalimantan.   Figure 11

This is the smallest drongo, slender and glossy black with a deeply forked tail. It is a common forest bird of the lowlands and lower hills. In the mountains it is replaced by the ASHY DRONGO *Dicrurus leucophaeus* (28 cm), which is grey in colour, the shade of grey varying according to sub-species. The upper-parts are a darker grey, and commonly there is a white patch on the cheek.

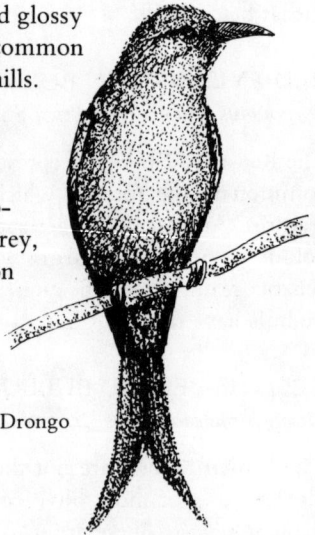

11. Bronzed Drongo

GREATER RACKET-TAILED DRONGO
Dicrurus paradiseus (33 cm, plus up to 30 cm of tail rackets).
Sumatra and Kalimantan. Figure 12

This common drongo is usually
unmistakable with its enormously
lengthened outer tail feathers, each
bearing a spoon-shaped racket at the
tip. However, in many birds, these
racket feathers are missing, either
through abnormal development or
when the bird is in moult. At such
times, it is not so easily distinguished
from the other black species. The almost
identical LESSER RACKET-TAILED
DRONGO *Dicrurus remifer* replaces
it in the mountains of Sumatra, but
this species is absent from Kalimantan.

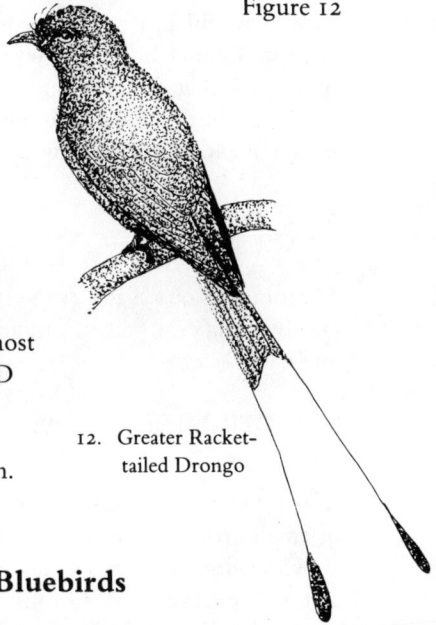

12. Greater Racket-
tailed Drongo

Orioles and Bluebirds

DARK-THROATED ORIOLE
Oriolus xanthonotus (19 cm). Sumatra and Kalimantan. Plate 17a

This rather small oriole is a common bird of forests in the lowlands
and lower hills. As it remains mostly in the canopy, it is difficult to
see, but hardly a day passes in the forest without hearing it at least
once. The song is typical for an oriole, consisting of a short, fluty
phrase. The head region is black, and only the upper-parts are
yellow, the belly being whitish with dark streaks. The female is quite
different, olive-green above and white streaked below, and lacking
the black head.

The BLACK-NAPED ORIOLE *O. chinensis* (27 cm) is quite
common in more open country in Sumatra. It is wholly golden
yellow, except for a black stripe through the face and round the nape,
and the black primary wing feathers and tail. It is very rare in
Kalimantan.

ASIAN FAIRY BLUEBIRD

Irena puella (25 cm). Sumatra and Kalimantan.   Plate 17b

This beautiful bird is quite common in lowland forests. The male, with its shining blue and velvety black, cannot be mistaken for any other bird. The female is more slaty blue with dark wings. Although a canopy bird, it is active and readily seen, and quite noisy with distinctive loud and mellow notes.

Crows

The omnivorous black crows are familiar to most people, but the family includes a range of quite interesting and sometimes colourful birds.

SLENDER-BILLED CROW

Corvus enca (46 cm). Sumatra and Kalimantan.   Figure 13

This is the common crow of woodland and forests, often seen in pairs or small groups, and occasionally in quite large flocks. Characteristic are its shallow wing-beats, and its rather nasal and high-pitched croaking calls. In more open country in Sumatra, it is replaced by the LARGE-BILLED CROW *C. macrorhynchos* (51 cm), which is usually seen in twos and threes, and has a deeper croak. The difference in the size of bill, from which the two species are named, is not a very distinctive field identification feature.

13. Slender-billed Crow

CRESTED JAY

Platylophus galericulatus (30 cm). Sumatra and
   Kalimantan.   Plate 17e

Although this very distinctive-looking bird is not often seen, as it stays deep in the forest, it is quite common. Its presence is often revealed by its call, which consists of a harsh rattle, readily recognized once known. It is a frequent participant in the mixed feeding flocks that are such a common feature of tropical forests.

More readily seen is the BLACK MAGPIE *Platysmurus leucopterus* (40 cm), which prefers forest edges and more open woodlands, including neglected rubber estates. This is a black bird with a moderately long tail and a small crest. Birds in Sumatra, but not in Kalimantan, have a white patch on the wing. It has a variety of peculiar calls, including bell-like notes uttered singly or in short ascending runs, as well as more discordant squawks

SUNDA TREEPIE
Dendrocitta occipitalis (43 cm). Sumatra and Kalimantan.　　Plate 17c

Although often named Malaysian Treepie, this bird does not occur in the Malay Peninsula, being endemic to Sumatra and Borneo, though it has a close relative on mainland Asia. It is quite common on the mountains of both islands, mainly on the edges of woodland, and is an attractive bird with its striking long tail and loud, varied calls, including bell-like notes and harsh chatters.

GREEN MAGPIE
Cissa chinensis (38 cm). Sumatra and Kalimantan.　　Plate 17d

One would expect these beautiful birds to be easy to see, but they are shy birds of the under-storey of montane forests in Sumatra, often only detected when their loud calls give them away. Generally an observer will get only short glimpses as single birds of a party fly from thicket to thicket. It has been reported on the northern borders of East Kalimantan but both this species and the very similar SHORT-TAILED GREEN MAGPIE *C. thalassina* occur in the mountains of northern Borneo, and should be looked for in Kalimantan.

Tits and Nuthatches

VELVET-FRONTED NUTHATCH
Sitta frontalis (13 cm). Sumatra and Kalimantan.　　Plate 17f

Small, compact, short-tailed birds, nuthatches clamber about the trunks and branches of trees in woodland, and have the unique ability to climb vertically down the trunk, facing down. This is a bird of lowland forests, which has all whitish under-parts, and a red bill. In the mountains of Sumatra (but not Kalimantan), it is replaced by the BLUE NUTHATCH *S. azurea*, which differs in having bluish-black upper-parts and belly, only the throat and breast being white; the bill is light blue.

GREAT TIT
Parus major (13 cm). Sumatra and Kalimantan.   Figure 14

In the Sundanese region, the Great Tit is a grey-and-black version of its yellow-and-black cousin of Europe, with prominent white cheeks in a black head. It has a curious distribution, occurring quite widely in the highlands of Sumatra, especially in the north, where it is common around Brestagi and Prapat. Elsewhere, it is a scarce bird of coastal woodlands and mangroves, which are its usual habitats in the Malay Peninsula and northern Borneo; it has recently been found along the desolate south coast of Kalimantan, in mangroves and adjacent tidal area settlements. Its varied calls, often rendered as 'pi-chu, pi-chu', commonly reveal its presence.

14. Great Tit

9. (a) Red-naped Trogon, p. 25. (b) Scarlet-rumped Trogon, p. 26.
(c) Stork-billed Kingfisher, p. 27. (d) White-throated Kingfisher,
p. 27. (e) Common Kingfisher, p. 27. (f) Oriental Dwarf
Kingfisher, p. 26.

10. (a) Blue-throated Bee-eater, p. 28. (b) Red-bearded Bee-eater, p. 28. (c) Dollarbird, p. 29. (d) Rhinoceros Hornbill, p. 29. (e) Wrinkled Hornbill, p. 30. (f) Helmeted Hornbill, p. 31.

11. (a) Red-crowned Barbet, p. 32. (b) Red-throated Barbet, p. 33.
    (c) Gold-whiskered Barbet, p. 33. (d) Yellow-crowned Barbet, p. 33.
    (e) Blue-eared Barbet, p. 32.

12. (a) Coppersmith Barbet, p. 33.   (b) Black-browed Barbet, p. 34.
(c) Fire-tufted Barbet, p. 34.   (d) Mountain Barbet, p. 34.
(e) Brown-capped Woodpecker, p. 35.   (f) Grey-and-Buff
Woodpecker, p. 35.

13. (a) Checker-throated Woodpecker, p. 36. (b) Maroon Woodpecker, p. 37. (c) Lesser Yellow-naped Woodpecker, p. 36. (d) Common Golden-backed Woodpecker, p. 37. (e) White-bellied Woodpecker, p. 37.

14. (a) Black-and-Red Broadbill, p. 39. (b) Black-and-Yellow
Broadbill, p. 38. (c) Green Broadbill, p. 39. (d) Banded
Broadbill, p. 38. (e) Dusky Broadbill, p. 39.

15. (a) Garnet Pitta, p. 40.   (b) Lesser Blue-winged Pitta, p. 41.
    (c) Blue-headed Pitta, p. 41.   (d) Fiery Minivet, p. 44.
    (e) Mountain Minivet, p. 45.   (f) Common Iora, p. 45.
    (g) Greater Green Leafbird, p. 45.

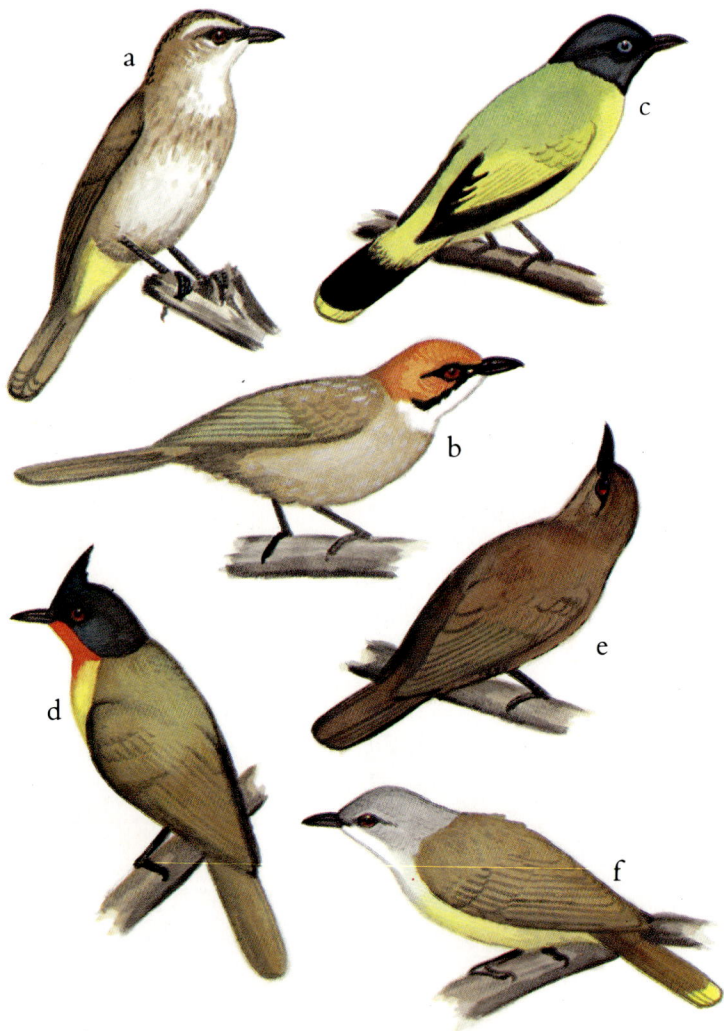

16. (a) Yellow-vented Bulbul, p. 46.   (b) Straw-headed Bulbul, p. 46.
     (c) Black-headed Bulbul, p. 47.   (d) Ruby-throated Bulbul, p. 47.
     (e) Red-eyed Brown Bulbul, p. 47.   (f) Yellow-bellied Bulbul, p. 47.

Babblers

This is the second big family of small forest birds, with over 40 species. While some are easy to identify from their plumage, others lack distinctive features and present as much difficulty as the brown bulbuls. However, nearly all have identifiable songs or calls, although it requires intensive study to learn these. Published tape recordings of babbler songs may help considerably, though the learner should always endeavour to confirm these for himself because the experts sometimes make mistakes! Identification is not assisted by the babblers' skulking habits, or the tendency of some groups to associate in mixed feeding parties. Only some of the more distinctive and common species are described here.

BLACK-CAPPED BABBLER
Pellorneum capistratum (18 cm). Sumatra and Kalimantan.   Plate 18a

This short-tailed babbler is common in the ground layer of dryland forest in the plains and lower hills. A very skulking bird, it walks stealthily about on the ground, but a quiet bird-watcher can usually obtain a view by following up the frequently uttered calls, and as Plate 18 shows, the plumage is distinctive. There appears to be a black stripe along the crown. The typical call is a soft, slightly disyllabic rising whistle, 'pheet', but there are other calls, a common one being a double whistled 'tip-ee', the second note usually higher, though not always.

ABBOTT'S BABBLER
Trichastoma abbotti (16 cm). Sumatra and Kalimantan.   Plate 18b

The babblers of this genus are all rather similar, being various shades of brown; the tails are rather short. They are skulking birds in the undergrowth of lowland primary and secondary forests, although Abbott's Babbler is more commonly found in thickets of secondary growth on the forest edge. It has a loud and monotonous song, typically of three sliding or disyllabic whistles, the second pitched lower though the whole phrase rises in pitch.

The WHITE-CHESTED BABBLER *T. rostratum* favours dense waterside growth along rivers and streams, in swamps, and at the inland margin of mangroves. Although very similar to that of Abbott's Babbler, the song is rather more musical and cheerful, rising more steeply in pitch with three or four disyllabic notes.

LARGER RED-HEADED BABBLER
Malacopteron magnum (18 cm). Sumatra and Kalimantan.   Plate 18c

The babblers of this genus are usually found in family parties in the undergrowth or lower storey of dryland forests. This is one of two species that have a dark rufous forecrown. The song is quite musical, consisting of clear whistled notes descending the scale, but sometimes rising at the start and end of the cadence, forming quite a complex sequence.

Very similar in appearance is the SMALLER RED-HEADED BABBLER *M. cinereum* whose song, however, consists simply of five or six thin notes rising the scale.

A favourite and rather startling song of stream-side campsites in lowland forest is that of the SOOTY-CAPPED BABBLER *M. affine*, which heralds the dawn with its extraordinary tuneless whistle of single notes, pitched higher and lower in a sort of sequence but never quite hitting the top notes. This has been likened to the aimless whistling of an errand boy, and in Kalimantan, the song seems to be especially pronounced in its tuneless simplicity.

LARGE WREN-BABBLER
Napothera macrodactyla (19 cm). Sumatra.   Plate 18d

The wren-babblers are so named because of their rather squat bodies with very short tails. Typically, they have rather mottled or scaly plumage, and they live close to the ground in a variety of forest habitats. The Large Wren-Babbler is a rather uncommon bird of lowland forests. The song is usually a succession of complex phrases of loud, fluty and often slurred, whistled notes, the phrases repeated with variations introduced one at a time.

In Kalimantan, this species is replaced by the endemic BLACK-

THROATED WREN-BABBLER *N. atrigularis*, which differs in having a black throat, although in other respects, including the song, it is very similar. It is found in wet forest, including peat swamps, as well as in the mountains.

BLACK-THROATED BABBLER

Stachyris nigricollis (15 cm). Sumatra and Kalimantan. Plate 18e

The babblers of the genus *Stachyris* are quite distinctive in both plumage and song, and they are less prone than the other babblers to mixed flocking behaviour. The Black-throated is common in the lowlands, and has a striking song of clear, deliberate whistles on one pitch, a dozen or more notes of which the first one or two are emphasized. Typical of babblers, this song is frequently accompanied by low churring notes from the mate.

The CHESTNUT-WINGED BABBLER *S. erythroptera* is a very common bird of the undergrowth. It has a whistled trill on one pitch, just too fast for a human to imitate, although attempts will usually incite a response. (The Pearl-cheeked Babbler *S. melanothorax* of Java and Bali has an almost identical song.) Distinctive features are the chestnut wings and dark foreparts, with a blue eye-ring.

The larger CHESTNUT-RUMPED BABBLER *S. maculata* also has a blue eye-ring, but with a black throat, black spots on a white breast, and a chestnut rump. The song is not easy to describe, consisting of clear di- or tri-syllabic phrases repeated in series, accompanied by trills from the mate. The song is easy to imitate, and the author has succeeded in whistling in this bird over a range of at least a hundred metres. It is quite common in lowland forests.

The GOLDEN BABBLER *S. chrysaea* (13 cm) is a smaller member of the group. Bright yellowish in colour, with a dark striped crown and a dark patch in front of the eye, it is very common in the mountains of Sumatra. It has a short whistled trill on one note, thinner and higher-pitched than that of the Black-throated, and characterized by the emphasized first note, and usually a brief pause after the first or second.

FLUFFY-BACKED TIT-BABBLER
Macronous ptilosus (16 cm). Sumatra and Kalimantan.   Plate 18f

The Fluffy-backed Tit-Babbler is named from the white-shafted spines on the back, although these can rarely be seen well. Its most striking feature is its call. The male gives three or four hollow whistles, 'poop, poop, poop', which draws an immediate outburst of loud squawking from the female, 'kedeke-kreuer, kedeke-kreuer'. When suspecting this bird to be present, the author has on occasion startled his colleagues in the forest by imitating the male's whistle, which immediately elicited the expected squawking response from birds all round the party. It is typically found in the secondary scrub and fern tangles of old *ladang* or the margins of logging roads in lowland forest.

Another babbler of secondary growth is the STRIPED TIT-BABBLER *Macronous gularis* (13 cm), marked by a rufous crown and black streaks on the yellow breast. The call is a prominent and repetitive 'chonk, chonk, chonk', heard commonly in this habitat, and usually accompanied by chuckling notes from other members of the party.

CHESTNUT-BACKED SCIMITAR BABBLER
Pomatorhinus montanus (20 cm). Sumatra and
   Kalimantan.   Plate 19b

This distinctive babbler occurs in quite noisy and conspicuous parties in the montane forests of Sumatra, where it gives a variety of cheerful calls. In contrast, in the lowland forests of both Sumatra and Kalimantan, where it seems to be uncommcn, it is a far more secretive bird, most often identified by its call, which here consists merely of a mellow triple hoot.

WHITE-CRESTED LAUGHING-THRUSH
Garrulax leucolophus (30 cm). Sumatra.   Plate 19c

This is one of the author's favourite Sumatran birds, and it is also very popular as a cage-bird. The charm of laughing-thrushes lies in their light, hopping, and inquisitive gait, and their outbursts of loud and cheerful noise. The climax of a flock's vocalizations has been described as a 'chorus of diabolical cackling laughter', perhaps a harsh description for so charming a bird. It is common in the lower storey of forest and forest edge habitats through the mountains of Sumatra, and is treated as a sub-species of a rather different bird found in mainland Asia. (Note that in Sumatran birds, the white foreparts are more confined to the head region than shown in Plate 19, and the black mask through the eye may be very narrow.)

CHESTNUT-CAPPED LAUGHING-THRUSH
Garrulax mitratus (25 cm). Sumatra and Kalimantan.   Plate 19d

The Chestnut-capped Laughing-Thrush occurs commonly in flocks in the montane forests of Sumatra, and along the northern borders of Kalimantan. It is another attractive bird, with a prominent white stripe in the wing, and an orange bill and legs. The calls consist of three or four rich, thrush-like notes.

SILVER-EARED MESIA
Leiothrix argentauris (19 cm). Sumatra.   Plate 19e

This very colourful babbler is quite common in the mountains of western and northern Sumatra. The plumage shown in Plate 19 is both beautiful and striking. It is found in active and quite noisy parties in forest edge habitats at moderate altitudes. The calls are varied but most often heard is a short, high, and shrill chatter.

WHITE-BROWED SHRIKE-BABBLER
Pteruthius flaviscapis (14 cm). Sumatra and Kalimantan.   Plate 19f

This is another colourful babbler of the mountain forests, found in the middle and upper storeys. The female is less striking, with a grey head, faint eye-stripe, and lack of pronounced colours in the wing. It is quite common, occurring in Sumatra up to the moss forest level. It is also common in northern Borneo, and a recent record from Gunung Niut is probably the first for Kalimantan.

The LONG-TAILED SIBIA *Heterophasia picoides* (30 cm) is also found in the montane forests of Sumatra (but not Kalimantan). It is a long-tailed bird, grey or brownish-grey, with a white patch in the wing, and a paler tip to the blackish tail. It occurs in flocks in the upper canopy.

MALAYSIAN RAIL BABBLER
Eupetes macrocerus (29 cm). Sumatra and Kalimantan. Plate 19a

This secretive bird of the forest floor was once thought to be very rare, until ornithologists learnt its call, since when it has been found to be quite common. This is illustrative of how secretive some forest birds are, and of the importance of learning their calls. It is difficult to do justice to the beauty of its plumage, as it walks purposely about on the ground, usually keeping just out of sight. However, imitations of its whistle will usually produce a brief sighting, when one may even see it flicking its wing as it walks, uttering soft 'tuk-tuk' notes. The call is a plaintive, soft, monotone whistle of about a second or longer, which is easy to imitate. Be careful, however, as it takes an expert ear to distinguish this from the almost identical call of the Garnet Pitta (see p. 40). The pitta's call is lower, shorter, and sharper, with a slight inflexion. The Rail Babbler is found in dryland forest in the plains and at least the lower mountains.

Thrushes

The thrushes form a large family of insectivorous birds, mostly living in woodland, that includes the true thrushes, robins, and forktails. They often have quite striking plumage, and some are renowned songsters.

WHITE-RUMPED SHAMA
Copsychus malabaricus (27 cm). Sumatra and Kalimantan. Plate 20a

Not only does the Shama have beautiful plumage, it is also the forest's best songbird. The song consists of rich, liquid, warbled phrases, which can be quite varied, although singing ability seems to vary between individuals. However, it is a skulking bird of the undergrowth and not very easy to see, except for its white back as it flies into cover, where it will sit giving clicking notes in alarm, flicking its tail in accompaniment. The female has broadly the same plumage but is somewhat duller. It is found in the lowlands and hills throughout the region. In Sabah, this bird has a white crown, and is separated here as a full species (*stricklandi*); it also occurs in the north-east corner of Kalimantan.

MAGPIE ROBIN
Copsychus saularis (22 cm). Sumatra and Kalimantan.   Plate 20b

The black-and-white Magpie Robin is common in more open wood-land and around villages. It is usually seen on the ground or lower branches, where it frequently cocks its tail. The song consists of short, sweet phrases that lack the fullness of the Shama's.

CHESTNUT-NAPED FORKTAIL
Enicurus ruficapillus (20 cm). Sumatra and Kalimantan.   Plate 20f

The forktails derive their name from their scissor-like, deeply forked, black-and-white barred tails. This species is the most colourful, with its chestnut nape and mantle. It lives along the clear, swifter, and more rocky streams of the forests, especially in the hills, extending into the adjacent lowlands. Forktails in general are quite wary, and most often seen in flight along the river, but their shrill, piping calls can be heard above the noise of the water. They are very active, dainty feeders, rather after the fashion of wagtails.

The WHITE-CROWNED FORKTAIL *E. leschenaulti* (20 cm) is found in both Sumatra and Kalimantan, along upland streams and also in wet lowland forests. It is black and white, with a white crown.

SUMATRAN WHISTLING THRUSH
Myiophoneus melanurus (23 cm). Sumatra.   Plate 20e

The Sumatran mountains have no less than three whistling thrushes, so they need to be looked at carefully. This species is endemic to Sumatra, where it is quite common near the ground in forest on the higher mountains. It is dark blue with bright spangles in the plumage, and shining blue eye-brows and shoulders. The bill is small and black, distinguishing it from the BLUE WHISTLING THRUSH *M. caeruleus* (32 cm) which has a yellow bill; this occurs near rocky watercourses in the hills.

Warblers

The warblers form a large family of rather small, thin-billed insect-eaters, mostly rather dull-coloured. They are very active birds in the foliage, and named from their warbling songs, though most Indonesian species are not good singers.

BLACK-NECKED TAILORBIRD
Orthotomus atrogularis (11 cm). Sumatra and Kalimantan.   Plate 20c

The tailorbirds are small, long-tailed warblers that are quite noisy but secretive birds of the undergrowth and scrub. They are named from their remarkable woven, domed nests. Most species are dull greenish above, and paler below. This species has a pronounced chestnut cap, green tail, and yellow under tail coverts; there is a black throat patch, though this may not always be obvious. The length of tail varies according to sex and season, but it is usually longer than shown in Plate 20. It is a common bird of scrubby forest margins in the lowlands and the lower hills.

The RUFOUS-TAILED TAILORBIRD *O. sericeus* has both the crown and tail rufous, and the ASHY (or RED-HEADED) TAILORBIRD *O. sepium* has the whole head rufous, with a greenish tail. They occur on both islands, as does the MOUNTAIN TAILORBIRD *O. cucullatus*, which has a longer bill, long white eye-brow, and bright yellow belly, and a varied musical repertoire. The last-named species has only recently been discovered in Kalimantan, on Gunung Niut in the west.

YELLOW-BELLIED PRINIA

Prinia flaviventris (13 cm). Sumatra and Kalimantan.   Plate 20d

Superficially similar to tailorbirds, prinias have a more elongated
appearance, with rather longer, slightly graduated tails, and more
uniform plumage. The Yellow-bellied Prinia is olive above, with a
grey head, whitish throat, and buffy yellowish belly. It is common in
all forms of wet grasslands including rank ricefields, where its cheerful little 'seeleelak' call is a constant companion. It is found in the
highlands, at least to the altitude of Lake Toba.

In Sumatra, the BAR-WINGED PRINIA *P. familiaris* is common
in drier scrub in all types of open vegetation including town gardens.
A double whitish wing bar is prominent. It is endemic to Sumatra,
Java, and Bali.

Also in Sumatra, mainly in the hills and mountains, from about
500 to 2 000 m, is the HILL PRINIA *P. atrogularis* (18 cm), which is
distinguished by a particularly long, somewhat untidy tail, and a half-
length white eye-brow, set in a grey head. The legs are orange. The
call is a rising, almost disyllabic 'shweet'. It is found in dry scrub and
undergrowth.

Flycatchers

The flycatchers form a large family of woodland insect-eaters, with
some 27 species in Sumatra and 24 in Kalimantan, including one or
two endemics on both islands. In addition, there are a few migrant
species. They tend to be quiet and unobtrusive, and although some
are quite colourful, they are not always easy to identify. Those in the
genus *Rhinomyias* and most species in the genus *Muscicapa* are brown,
and it needs an expert to identify them. Many have songs, but these
tend to be short, rather thin phrases which sound alike. The greatest
variety of species is found in the hills.

MALAYSIAN BLUE FLYCATCHER
Cyornis turcosa (14 cm). Sumatra and Kalimantan.   Plate 21b

This is one of the group of flycatchers having quite striking blue upper-parts and rufous under-parts. In this species, the throat in the male is also blue. However, the female has the throat the same buffy rufous as the rest of the under-parts, and is very similar to the other blue flycatchers. Generally, this bird is found in pairs, so identification is not difficult. It is found in lowland primary forest, especially along rivers, where it is quite common, particularly in Kalimantan.

The MANGROVE BLUE FLYCATCHER *C. rufigastra* occurs in coastal areas and the HILL BLUE FLYCATCHER *C. banyumas* in the hills. In both species, the male has the throat the same rufous as the breast and belly, so they look like more strongly coloured versions of the female of the previous species. As there are at least two other species having a similar colour distribution, however, great care is needed in identification.

VERDITER FLYCATCHER
Muscicapa thalassina (16 cm). Sumatra and Kalimantan.   Plate 21d

Another group of flycatchers is mainly blue throughout. This species is greenish-blue, when viewed in a good light. There is a small black patch in front of the eyes. The female is similar but duller. It occurs in the hills, where it feeds by flying out to catch insects from an exposed perch.

The INDIGO FLYCATCHER *M. indigo* is a much darker blue, with a whitish belly, and a pale blue forehead above a dark face. It is common in the higher mountains of Sumatra, and although not yet reported from the Indonesian portion of Borneo, it probably occurs along the border mountains.

Once again, there are two or three other predominantly blue species in the hills, to confuse the beginner. In Sumatra, one of these is the LARGE NILTAVA *Niltava grandis*, a large flycatcher (22 cm) that might at first be mistaken for one of the blue thrushes. It is very dark blue, appearing blackish in poor light, with a shining blue crown, nape, and patch in the shoulders.

RUFOUS-CHESTED FLYCATCHER

Ficedula dumetoria (11 cm). Sumatra and Kalimantan.　　　Plate 21e

Yet another group of flycatchers has the upper-parts very dark grey to black. This species has a rufous breast, with a paler throat and belly, and white patches above the eye, in the wing, and sides to the base of the tail (not shown in Plate 21). It is a quiet bird of the under-storey of the lower hills, but it is quite tame and not difficult to watch. Although it is not very common, recent records from the coastal plain of Tanjung Puting in Kalimantan suggest that it is widely distributed.

In the higher mountains, though not yet reported from Indonesian Borneo, the SNOWY-BROWED FLYCATCHER *F. hyperythra* is dark bluish-grey above, with a short, thick white eye-brow in front of the eye.

A striking winter migrant to Sumatra, which occurs regularly in Way Kambas, is the YELLOW-RUMPED FLYCATCHER *F. zanthopygia*, a lovely bird with black upper-parts and yellow under-parts. The rump is also yellow, and there is a white eye-brow and patch in the wing.

GREY-HEADED FLYCATCHER

Culicicapa ceylonensis (13 cm). Sumatra and Kalimantan.　　　Plate 21f

The Grey-headed Flycatcher is common in the middle storey of the forest in the lowlands and hills. It has a warbler-like appearance and behaviour. The entire head region is grey, the upper-parts are olive, and the under-parts yellow. Its simple four-note call, alternately low and high, 'silly billy', often reveals its presence.

WHITE-THROATED FANTAIL

Rhipidura albicollis (18 cm). Sumatra and Kalimantan.　　　Plate 21g

As the name implies, the fantails have rather long tails which are frequently spread like a fan, exposing the white tips of the feathers. This is a common fantail in the mountains of Sumatra, and it has recently been discovered in Kalimantan on Gunung Niut. It is greyish-brown all over, except for a thin white stripe above the eye and a small white throat patch.

In the lowlands, it is replaced by the PIED FANTAIL *R. javanica*, which differs in having all the under-parts white, except for a dark breast band. It is a common bird of woodland margins, and even villages, favouring especially damp scrub, and it is not averse to coming into town gardens; it has a cheerful song of loud, short, warbled phrases.

BLACK-NAPED MONARCH

Hypothymis azurea (14 cm). Sumatra and Kalimantan.   Plate 21c

This is the commonest member of a group of flycatchers known as the monarchs. Plate 21 shows it to be a very distinctive bird, unmistakable if seen clearly. The female is slightly duller, and lacks the black markings. It is found in the middle storey of lowland forests and the lower hills, and its 'wit–wit–wit' call often reveals its presence.

ASIAN PARADISE FLYCATCHER

Terpsiphone paradisi (20 cm, plus 23 cm of elongated tail plumes in the male). Sumatra and Kalimantan.

The remarkable Asian Paradise Flycatcher has a call very similar to that of the previous species. The male is a white bird, with a black head region and primaries, and two white tail plumes up to 23 cm long. It is quite common in lowland forests, and unmistakable when seen well. Note, however, that the female is not white at all, being rufous, blacker around the head, and paler below, and lacking the elongated tail. Some males also occur in a rufous phase which is similar to that of the female.

MANGROVE WHISTLER

Pachycephala cinerea (15 cm). Sumatra and Kalimantan.   Plate 21a

Related to the flycatchers, this is a representative of a large Australian family. It is rather nondescript, greyish-brown, paler below, with a noticeably rounded head. The most distinctive feature is its varied song, which typically consists of a series of loud whistled notes on one pitch, followed by a single or double explosive note. (This should not be confused with the 'weeeeeeee-tu' of the Common Iora, see p. 45). It has, however, a wide range of phrases, including warbled versions, but the explosive notes usually give it away. It is not confined to mangroves, being equally at home in a variety of lowland forests, though usually in the thinner forests growing on poorer quality, sandy, or peaty soils, and in old rubber plantations.

Wagtails and Pipits

GREY WAGTAIL
Motacilla cinerea (19 cm). Sumatra and Kalimantan.   Plate 22a

The wagtails, so named from their long tails which are constantly wagged up and down as they feed on the ground, are common winter migrants to the Indonesian region, arriving in August or September. The Grey Wagtail has grey upper-parts and yellowish under-parts. In winter, the throat is white, though it becomes black in the breeding male. The call is a sharp, double 'chizik'. The normal habitat of Grey Wagtails is rocky mountain streams, but in winter they are usually found along tracks, especially in the hills.

In contrast, the YELLOW WAGTAIL *M. flava* is a common migrant to ricefields and open swamp areas, occurring in small flocks. It has a rather shorter tail and more olive-green upper-parts, and the call is a single wheezy 'tseep'.

RICHARD'S PIPIT *Anthus novaeseelandiae* (18 cm) is a ground bird of open grassy terrain. Brownish and slightly streaked, it is a slender bird with quite long legs, and, like the wagtails, it has white outer tail feathers. It is not uncommon in a variety of habitats, including airports, but it is easily overlooked by bird-watchers interested in seeking the more colourful birds of the forests.

Shrikes

BROWN SHRIKE

Lanius cristatus (19 cm). Sumatra and Kalimantan.   Plate 22b

Shrikes are medium-sized birds with strong heads and hooked bills used for catching large insects, which are sometimes impaled on thorns for storage. The Brown Shrike is a common winter visitor to lightly wooded country. Slightly rufous brown above and creamy below, it has a broad black band through the eye, and a white stripe above it and on the forehead. Young birds show some fine crescentic barring. The call is a harsh scolding chatter.

The LONG-TAILED SHRIKE *L. schach* (25 cm) is a larger, longer-tailed, and more colourful shrike that is a common resident throughout the open country of Sumatra, and at least the south-eastern quarter of Kalimantan. There are a number of sub-species, which vary according to the amount of black on the crown, and one form, a winter migrant to East Kalimantan, has an all-black crown. Although the usual call notes are quite harsh, its short song is not unmusical.

Another bird of open country is the WHITE-BREASTED WOOD-SWALLOW *Artamus leucorhynchus* (18 cm). Wood-Swallows live in small parties in habitats ranging from woodland edges to open areas in towns, where they perch on any high structures such as radio masts and even buildings. This is a squat bird, which feeds on insects by making brief sorties on distinctively triangular wings. It has slaty grey upper-parts and breast, white under-parts, a pale rump, and a thick, grey bill.

Bristlehead

BORNEAN BRISTLEHEAD

Pityriasis gymnocephala (25 cm). Kalimantan.   Plate 22c

This remarkable Bornean endemic is believed to be related to the butcher-birds of Australia and New Guinea, and it is interesting to speculate on just how this species evolved in isolation in Borneo. It is a somewhat enigmatic bird, unpredictable in occurrence, and liable to turn up anywhere in forest—typically in the dry lowlands, but it also wanders into peat swamps and the hills.

It is found in loose parties of up to a dozen, appearing rather like large mynas, all black with heavy black bill, beautiful red head, orange on the crown, and red thighs. The tail is short and there is a bluish-white flash in the wings in flight. It is not shy, and can be watched readily as individual birds in the party fly from tree to tree with short, whirring flights, occasionally giving a variety of short chortling calls. It is a bird that seems to be chanced upon by accident—those who go out to look for it will rarely be rewarded.

Mynas and Starlings

HILL MYNA
Gracula religiosa (30 cm). Sumatra and Kalimantan.   Plate 22d

Also known as GRACKLE, or TALKING MYNA, this bird is famous for its seeming ability to talk, although talking pet birds are only imitating the sounds their owners have taught them. It is to be hoped the pet trade will now leave this bird alone, as the wild population has been decimated in Java and elsewhere. A large, bluish-black bird with heavy orange bill and yellow lappets, there is a white patch in the rounded wings, prominent in flight. Its most typical call is a piercing 'ti-ong'. It is not uncommon in twos and threes in both lowland and hill forest, where it keeps to the upper canopy.

PHILIPPINE GLOSSY STARLING
Aplonis panayensis (20 cm). Sumatra and Kalimantan.   Plate 22e

The Glossy Starling is a common bird, found especially in coastal regions, where large numbers may congregate to roost in shrill parties in selected places such as groves of coconut palms or mangroves. It is a sleek, glossy green bird with red eyes, although it looks black in flight. Young birds are whitish below with dark streaks.

Other starlings which may be seen in Sumatra are the ASIAN PIED STARLING *Sturnus contra* and WHITE-VENTED MYNA *Acridotheres javanicus* from Java, and the COMMON MYNA *A. tristis* from the Malay Peninsula. It is not known to what extent Man has assisted in the spread of these birds.

Sunbirds and Spiderhunters

Sunbirds are the very active little birds with long, slender, curved bills that feed on the nectar of flowering shrubs, and the tiny insects found around them. Sometimes they hover in front of the flower like a hummingbird, but only briefly. Although the males are often very colourful, and sometimes glossy, the coloured parts may just appear blackish in a poor light, so that identification can be difficult. The females are especially difficult, most of them being alike, olive above and yellowish below. The same ten species are found in both Sumatra and Kalimantan, together with six of the very long billed spiderhunters (plus one endemic species in Borneo).

RUBY-CHEEKED SUNBIRD
Anthreptes singalensis (11 cm). Sumatra and Kalimantan.   Plate 23c

This sunbird is common in the upper storey of woodlands, and sometimes gardens and coconut palms, but it is not always easy to see. In a good light, the upper-parts of the male are a brilliant metallic green, with ruby-coloured cheeks. Both sexes have a rufous throat and upper breast, and a yellow belly.

However, the BROWN-THROATED SUNBIRD *A. malacensis* (14 cm) is more often seen. It is very similar, differing mainly in having a brown throat. Together with the next species, it is common in a variety of open habitats.

OLIVE-BACKED SUNBIRD
Nectarinia jugularis (11 cm). Sumatra and Kalimantan.   Plate 23b

The male of this species is distinguished by its olive upper-parts. The throat is a metallic near-black, and the belly is bright yellow. All the under-parts of the female are yellow. Useful characters are the rather longer and thinner bill than in the Brown-throated, and particularly the white lateral feathers on the dark tail (not shown on Plate 23).

17. (a) Dark-throated Oriole, p. 49. (b) Asian Fairy Bluebird, p. 50.
(c) Sunda Treepie, p. 51. (d) Green Magpie, p. 51. (e) Crested
Jay, p. 50. (f) Velvet-fronted Nuthatch, p. 51.

18. (a) Black-capped Babbler, p. 53. (b) Abbott's Babbler, p. 53.
(c) Larger Red-headed Babbler, p. 54. (d) Large Wren-Babbler, p. 54.
(e) Black-throated Babbler, p. 55. (f) Fluffy-backed
Tit-Babbler, p. 56.

19. (a) Malaysian Rail Babbler, p. 58.   (b) Chestnut-backed
Babbler, p. 56.   (c) White-crested Laughing-Thrush, p. 56.
(d) Chestnut-capped Laughing-Thrush, p. 57.   (e) Silver-eared
Mesia, p. 57.   (f) White-browed Shrike-Babbler, p. 57.

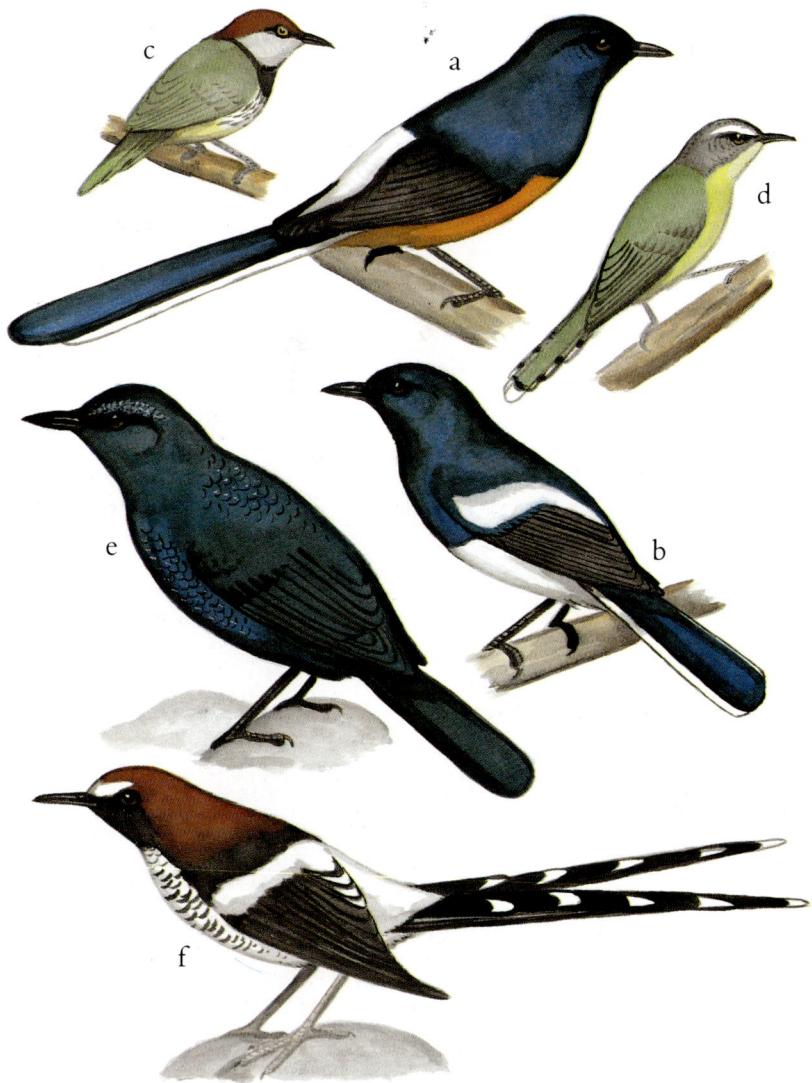

20. (a) White-rumped Shama, p. 58. (b) Magpie Robin, p. 59.
(c) Black-necked Tailorbird, p. 60. (d) Yellow-bellied Prinia, p. 61.
(e) Sumatran Whistling Thrush, p. 59. (f) Chestnut-naped
Forktail, p. 59.

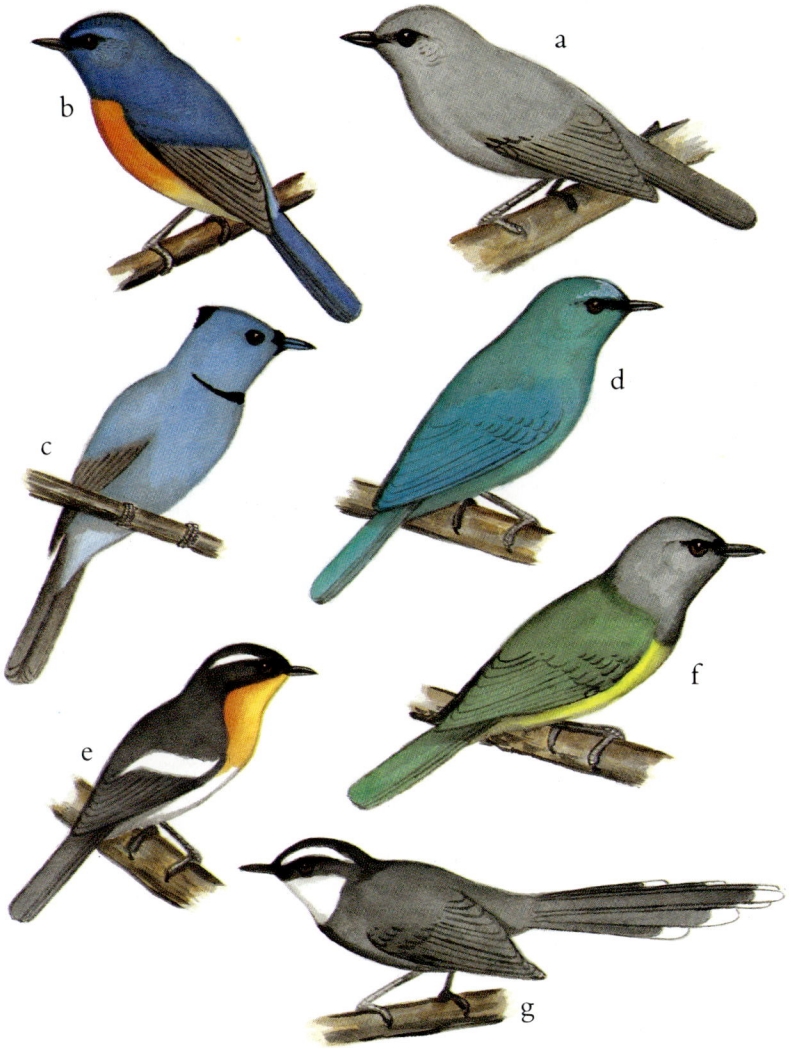

21. (a) Mangrove Whistler, p. 64.   (b) Malaysian Blue Flycatcher, p. 62.
(c) Black-naped Monarch, p. 64.   (d) Verditer Flycatcher, p. 62.
(e) Rufous-chested Flycatcher, p. 63.   (f) Grey-headed
Flycatcher, p. 63.   (g) White-throated Fantail, p. 63.

22. (a) Grey Wagtail, p. 65. (b) Brown Shrike, p. 66. (c) Bornean Bristlehead, p. 66. (d) Hill Myna, p. 67. (e) Philippine Glossy Starling, p. 67.

23. (a) Yellow-backed Sunbird, p. 69. (b) Olive-backed Sunbird, p. 68. (c) Ruby-cheeked Sunbird, p. 68. (d) Copper-throated Sunbird, p. 69. (e) Purple-naped Sunbird, p. 69. (f) Yellow-eared Spiderhunter, p. 69.

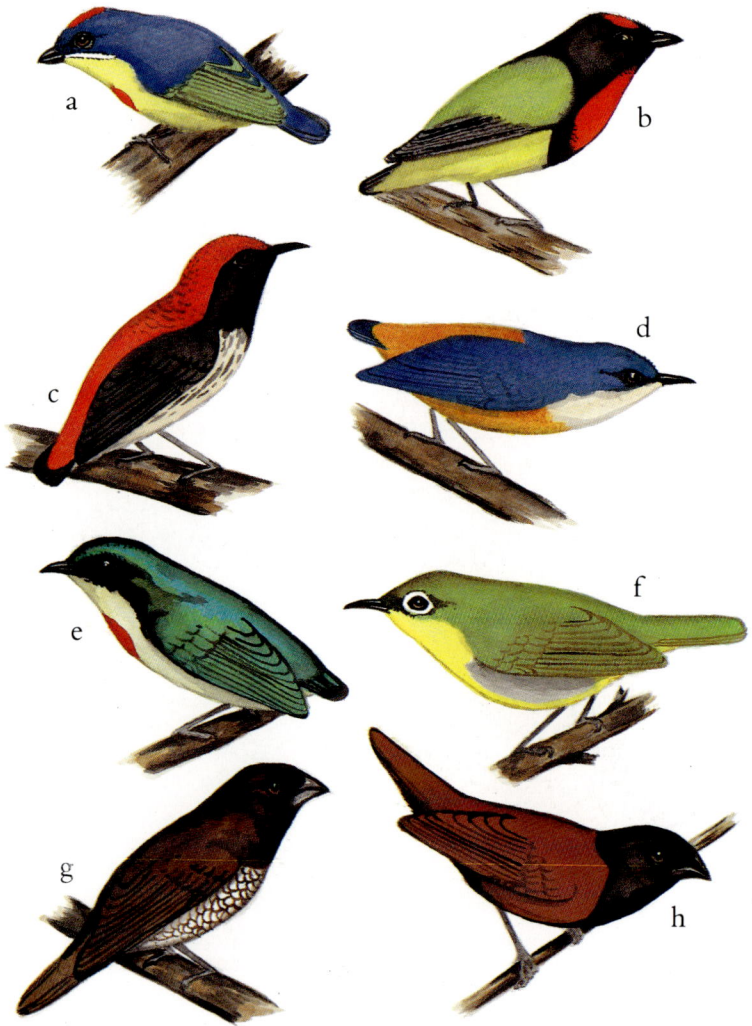

24. (a) Crimson-breasted Flowerpecker, p. 70. (b) Scarlet-breasted Flowerpecker, p. 70. (c) Scarlet-backed Flowerpecker, p. 71. (d) Orange-bellied Flowerpecker, p. 71. (e) Buff-bellied Flowerpecker, p. 71. (f) Oriental White-eye, p. 71. (g) Scaly-breasted Munia, p. 72. (h) Chestnut Munia, p. 73.

COPPER-THROATED SUNBIRD

Nectarinia calcostetha (14 cm). Sumatra and Kalimantan.   Plate 23d

This species occurs mainly in the coastal regions, in mangroves, casuarinas, coconut palms, and tall beach scrub. The male is almost wholly dark, with a variety of metallic colours, and with orange pectoral tufts beneath the wings, though these are often concealed. The tail is slightly graduated, and in the female it is tipped white.

PURPLE-NAPED SUNBIRD

Hypogramma hypogrammica (15 cm). Sumatra and
  Kalimantan.   Plate 23e

This large sunbird is quite distinctive with its streaked under-parts and, in the male, the metallic purple bar across the nape. There is also purple on the rump though this is less conspicuous. It is quite common in scrubby growth in the forest edge, often near rivers or swamps.

YELLOW-BACKED SUNBIRD

Aethopyga siparaja (11 cm). Sumatra and Kalimantan.   Plate 23a

The male sunbirds in this genus are particularly colourful, and in breeding plumage, they have elongated central tail feathers. This bird is crimson red with a glossy dark forecrown, moustachial stripe, and tail, and a yellow rump. The belly is dark grey. It occurs mainly in the lowlands and coastal areas in woodland edge habitats.

YELLOW-EARED SPIDERHUNTER

Arachnothera chrysogenys (18 cm). Sumatra and
  Kalimantan.   Plate 23f

It is difficult to select any one spiderhunter for description: they are all striking, with their enormously long, curved bills, and they are all partial to secondary scrub with banana and plantain growing in old *ladang* within and around the forest. Furthermore, they are particularly difficult to observe, being quite shy and usually seen only in fast, direct flight, when they look very alike, being green or yellowish-green. This species has quite pronounced yellow ear coverts, but this is not an exclusive feature.

Somewhat easier to identify is the LITTLE SPIDERHUNTER *A. longirostra* (15 cm), which has a whitish throat, yellow belly, and proportionally longer bill.

Flowerpeckers

Flowerpeckers are stockier than sunbirds, with short bills, but they have similarly active behaviour, constantly flitting from one flowering shrub to another on swift but erratic flight. They are particularly partial to the epiphytes growing on forest trees. Most of the males have bright colours, but without the glossy metallic hues that can be confusing in sunbirds. There are ten species in Sumatra, eleven in Kalimantan, of which eight are common to both, while two in Kalimantan are endemic. Only the more colourful and obtrusive species are described here.

CRIMSON-BREASTED FLOWERPECKER
Prionochilus percussus (10 cm). Sumatra and Kalimantan.   Plate 24a

The Crimson-breasted Flowerpecker is a common bird of the middle storey of primary and secondary forest in the lowlands and hills. As Plate 24 shows, the male is unmistakable, but in Kalimantan, a very similar endemic species is more common—the YELLOW-RUMPED FLOWERPECKER *P. xanthopygius*. This differs in having a yellow rump patch and lacking the white moustachial stripe.

SCARLET-BREASTED FLOWERPECKER
Prionochilus thoracicus (10 cm). Sumatra and Kalimantan.   Plate 24b

As Plate 24 shows, the Scarlet-breasted Flowerpecker is a particularly beautiful bird. It seems to be rare, however, though perhaps it lives more in the canopy and is therefore easily overlooked. Indeed, prior to two sightings in the 1970s (not yet repeated, at time of writing), it was unknown on mainland Sumatra.

SCARLET-BACKED FLOWERPECKER
Dicaeum cruentatum (9 cm). Sumatra and Kalimantan.   Plate 24c

The male of this common species of forests and wooded areas is distinctive with its black body, red crown, back and rump, and whitish streak down the centre of the under-parts. It is closely related to the SCARLET-HEADED FLOWERPECKER *D. trochileum* of Java, which differs in having the whole head and foreparts red. The latter bird occurs in the Banjarmasin region of Kalimantan, and has recently been discovered in southern Lampung.

ORANGE-BELLIED FLOWERPECKER
Dicaeum trigonostigma (9 cm). Sumatra and Kalimantan.   Plate 24d

Another distinctive species is the Orange-bellied, with its orange–yellow back, lower breast and belly, pale grey throat, and dark blue–grey crown, mantle and wings. It is a common forest and woodland bird of the lowlands and lower mountains.

BUFF-BELLIED FLOWERPECKER
Dicaeum ignipectus (9 cm). Sumatra.   Plate 24e

The Buff-bellied is the montane flowerpecker of Sumatra. It has blue–green upper-parts, a cinnamon patch on the breast, and a black line down the centre of the belly (this is not visible on Plate 24).

Its counterpart in the higher mountains of Kalimantan is the BLACK-SIDED FLOWERPECKER *D. monticolum*, which has dark slaty upper-parts and a prominent red throat and breast.

White-eyes

ORIENTAL WHITE-EYE
Zosterops palpebrosus (11 cm). Sumatra.   Plate 24f

The white-eyes are active birds of the canopy that behave rather like sunbirds or flowerpeckers, except that they usually occur in small, loose parties that call to each other incessantly, with cheeping notes rather like baby chickens. They are mainly olive in colour, some more green and others more yellow, and are so named because of the prominent white ring round the eye. They are not easy birds for the amateur to identify, and there are several species. The Oriental White-eye is quite common both in the coastal mangroves and adjacent woodlands of eastern Sumatra, and in the hills. In general, white-eyes are not prominent in the Kalimantan lowlands.

Sparrows, Weavers, and Munias

These families consist of the rather rounded, thick-billed seed-eaters, and include the munias that plunder the ricefields. Compared to the sub-tropical regions of Africa and Australia, they are not well represented in the Indonesian region, though the munias are common. The drab-coloured EURASIAN TREE SPARROW *Passer montanus* (15 cm) is quite common near human settlements throughout Sumatra, although it is quite a recent colonist in Kalimantan, where it is still rather local.

There are no weavers in Kalimantan, and the only species in Sumatra is the BAYA WEAVER *Ploceus philippinus* (14 cm), which nests in colonies in trees close to ricefields; their nests have long, suspended entrance spouts. In the breeding season, the male has a bright yellow crown, black facial mask, and pale, unstreaked underparts.

SCALY-BREASTED MUNIA
Lonchura punctulata (11 cm). Sumatra and Kalimantan.　　　　Plate 24g

One of the common munias in Sumatra is the Scaly-breasted or Spotted Munia, so named from the heavy crescentic black markings on the under-parts. The rest of the plumage is brown, becoming reddish around the bill. Although familiar, in 'cheeping' flocks in open scrub and grasslands as well as ricefields, they are quite wary, and quick to take flight. In Kalimantan, they have been discovered only recently, along the edge of the Lower Barito swamps north of Banjarmasin.

CHESTNUT MUNIA
Lonchura malacca (11 cm). Sumatra and Kalimantan.   Plate 24h

This munia is distinctive with its chestnut plumage, black head and foreparts, and bluish bill. In Sumatra, it often associates with the WHITE-HEADED MUNIA *L. maja*, which is chestnut with a whitish head. The latter bird is not found in Kalimantan.

DUSKY MUNIA
Lonchura fuscans (10 cm). Kalimantan.   Figure 15

The book closes with a singularly dull-coloured bird which is an abundant endemic in Kalimantan. It is brownish-black throughout, with a horn-coloured bill and pale bluish feet. It is seen in small, shy flocks which do considerable damage to farmers' crops throughout the island.

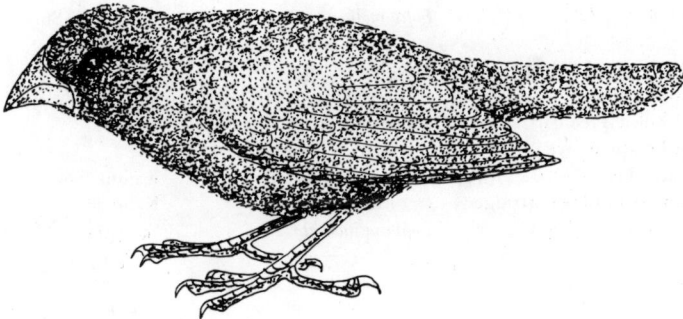

15. Dusky Munia

Appendix

A Checklist of Sundanese Endemics in Indonesia

THE following list tabulates all those species that are mainly endemic in the Sundanese region of Indonesia, which comprises Sumatra, Kalimantan, Java, and Bali, and most of their offshore islands. In this context, the Indonesian name of Kalimantan is used in the sense of the whole of Borneo, inclusive of northern Borneo (the Malaysian states of Sarawak and Sabah, and Negara Brunei Darussalam). An annotation of 'N. Borneo' is used for birds that do not occur in Indonesian Kalimantan.

This list is as definitive as possible, but taxonomists will always disagree on species definitions. Some of the birds listed—for example, the White-crowned Shama—may be sub-species of more widely distributed birds.

Mountain Serpent-Eagle	*Spilornis kinabaluensis*	N. Borneo
Javan Hawk-Eagle	*Spizaetus bartelsi*	Java
Spotted Kestrel	*Falco moluccensis*	Java, Bali, Sulawesi, Maluku
White-fronted Falconet	*Microhierax latifrons*	N. Borneo
Javan Partridge	*Arborophila javanica*	Java
Red-billed Tree Partridge	*Arborophila rubrirostris*	Sumatra
Red-breasted Tree Partridge	*Arborophila hyperythra*	Kalimantan
Crimson-headed Partridge	*Haematortyx sanguiniceps*	Kalimantan
Salvadori's Pheasant	*Lophura inornata*	Sumatra (*L. i. inornata* in western Sumatra, *L. i. hoogerwerfi* in northern Sumatra)
Bulwer's Pheasant	*Lophura bulweri*	Kalimantan
Green Junglefowl	*Gallus varius*	Java, Bali, Nusatenggara
Bronze-tailed Peacock-Pheasant	*Polyplectron chalcurum*	Sumatra
Bornean Peacock-Pheasant	*Polyplectron schleiermacheri*	Kalimantan
Javan Wattled Lapwing	*Vanellus macropterus*	Java (extinct)
Javan Sand Plover	*Charadrius javanicus*	Java, Bali
Indonesian Woodcock	*Scolopax saturata*	Sumatra, Java
Pin-tailed Green Pigeon	*Treron oxyura*	Sumatra, Java

Grey-headed Green Pigeon	*Treron griseicauda*	Java, Bali, Sulawesi, Sula Is.
Black-backed Fruit Dove	*Ptilinopus cinctus*	Bali, Nusatenggara
Pink-necked Fruit Dove	*Ptilinopus porphyreus*	Sumatra, Java, Bali
Dark-backed Imperial Pigeon	*Ducula lacernulata*	Java, Bali, W. Nusatenggara
Javan Hanging Parrot	*Loriculus pusillus*	Java, Bali
Javan Coucal	*Centropus nigrorufus*	Java
Sunda (Malay) Ground Coucal	*Carpococcyx radiceus*	Sumatra, Kalimantan
Stresemann's Mountain Scops Owl	*Otus stresemanni*	Sumatra
Simeuleuwe Scops Owl	*Otus umbra*	Simeuleuwe
Mentawai Scops Owl	*Otus mentawi*	Mentawai Is.
Enggano Scops Owl	*Otus enganensis*	Enggano
Javan Scops Owl	*Otus angelinae*	Java
Rajah's Scops Owl	*Otus brookii*	Sumatra, N. Borneo, Java
Javan Barred Owlet	*Glaucidium castanopterum*	Java, Bali
Sunda Frogmouth	*Batrachostomus cornutus*	Sumatra, Kalimantan, Kangean Is.
Dulit Frogmouth	*Batrachostomus harterti*	Kalimantan
Pale-headed Frogmouth	*Batrachostomus poliolophus*	Sumatra, Kalimantan
Bonaparte's Nightjar	*Caprimulgus concretus*	Sumatra, Kalimantan
Salvadori's Nightjar	*Caprimulgus pulchellus*	Sumatra, Java
Mossy-nest Swiftlet	*Aerodromus salangana*	Sumatra, N. Borneo, Java
Linchi Swiftlet	*Collocalia linchi*	Sumatra, Java, Bali, Lombok
Blue-tailed Trogon	*Harpactes reinwardtii*	Sumatra, Java
Whitehead's Trogon	*Harpactes whiteheadi*	N. Borneo
Javan Kingfisher	*Halcyon cyaniventris*	Java, Bali
Small Blue Kingfisher	*Alcedo coerulescens*	Sumatra, Java, Bali, W. Nusatenggara
Black-banded Barbet	*Megalaima javensis*	Java, Bali
Blue-crowned Barbet	*Megalaima armillaris*	Java, Bali
Brown-throated Barbet	*Megalaima corvina*	Java
Mountain Barbet	*Megalaima monticola*	Kalimantan
Black-throated Barbet	*Megalaima eximia*	Kalimantan
Golden-naped Barbet	*Megalaima pulcherrima*	N. Borneo
Hose's Broadbill	*Calyptomena hosei*	Kalimantan

Whitehead's Broadbill	*Calyptomena whiteheadi*	Kalimantan
Schneider's Pitta	*Pitta schneideri*	Sumatra
Blue-banded Pitta	*Pitta arquata*	Kalimantan
Black-and-Scarlet Pitta	*Pitta venusta*	Sumatra, N. Borneo
Blue-headed Pitta	*Pitta baudi*	Kalimantan
Black-faced Cuckoo-Shrike	*Coracina larvata*	Sumatra, N. Borneo, Java
Sunda Minivet	*Pericrocotus miniatus*	Sumatra, Java
Blue-masked Leafbird	*Chloropsis venusta*	Sumatra
Striated Green Bulbul	*Pycnonotus leucogrammicus*	Sumatra
Olive-crowned Bulbul	*Pycnonotus tympanistrigus*	Sumatra
Ruby-throated Bulbul	*Pycnonotus dispar*	Sumatra, Java
Blue-wattled Bulbul	*Pycnonotus nieuwenhuisii*	Sumatra, Kalimantan
Orange-spotted Bulbul	*Pycnonotus bimaculatus*	Sumatra, Java, Bali
Hook-billed Bulbul	*Setornis criniger*	Sumatra, Kalimantan
Streaked Mountain Bulbul	*Hypsipetes virescens*	Sumatra, Java
Black Oriole	*Oriolus hosei*	Kalimantan
Sunda Treepie	*Dendrocitta occipitalis*	Sumatra, Kalimantan
Pygmy Tit	*Psaltria exilis*	Java
Temminck's Babbler	*Trichastoma pyrrogenys*	Sumatra
Black-browed Babbler	*Trichastoma perspicillatum (vanderbilti)*	Sumatra, Kalimantan
Sumatran Large Wren-Babbler	*Napothera rufipectus*	Sumatra
Black-throated Wren-Babbler	*Napothera atrigularis*	Kalimantan
Mountain Wren-Babbler	*Napothera crassa*	Kalimantan
Bornean Wren-Babbler	*Ptilocichla leucogrammica*	Kalimantan
White-breasted Babbler	*Stachyris grammiceps*	Java
White-collared Babbler	*Stachyris thoracica*	Java
Pearl-cheeked Babbler	*Stachyris melanothorax*	Java, Bali
Javan Tit-Babbler	*Macronous flavicollis*	Java
Red-fronted Laughing-Thrush	*Garrulax rufifrons*	Java
Grey-and-Brown Laughing-Thrush	*Garrulax palliatus*	Sumatra, Kalimantan
Javan Fulvetta	*Alcippe pyrrhoptera*	Java
Chestnut-crested Babbler	*Yuhina everetti*	Kalimantan
Spotted Crocias	*Crocias albonotatus*	Java
White-crowned Shama	*Copsychus stricklandi*	Kalimantan
Sunda Blue Robin	*Cinclidium diana*	Sumatra, Java

Lesser Forktail	*Enicurus velatus*	Sumatra, Java
Javan Cochoa	*Cochoa azurea*	Java
Sumatran Cochoa	*Cochoa beccarii*	Sumatra
Sumatran Whistling Thrush	*Myiophoneus melanurus*	Sumatra
Sunda Whistling Thrush	*Myiophoneus glaucinus*	Sumatra, Kalimantan, Java
Everett's Thrush	*Zoothera everetti*	N. Borneo
Black-breasted Thrush (Triller)	*Chlamydochaera jefferyi*	Kalimantan
Sunda Flycatcher-Warbler	*Seicercus grammiceps*	Sumatra, Java
Short-tailed Bush Warbler	*Cettia whiteheadi*	Kalimantan
Kinabalu Friendly Warbler	*Bradypterus accentor*	N. Borneo
Bar-winged Prinia	*Prinia familiaris*	Sumatra, Java, Bali
Javan Tesia	*Tesia superciliaris*	Java
Indigo Flycatcher	*Muscicapa indigo*	Sumatra, N. Borneo, Java
Rueck's Blue Flycatcher	*Cyornis ruecki*	? Sumatra
Large-billed Blue Flycatcher	*Cyornis caerulata*	Sumatra, Kalimantan
Bornean Blue Flycatcher	*Cyornis superba*	Kalimantan
Red-tailed Fantail	*Rhipidura phoenicura*	Java
White-bellied Fantail	*Rhipidura euryura*	Java
Bornean Mountain Whistler	*Pachycephala hypoxantha*	Kalimantan
Black-winged Starling	*Sturnus melanopterus*	Java, Bali (introduced in Singapore)
Bali White Myna	*Leucopsar rothschildi*	Bali
Bornean Bristlehead	*Pityriasis gymnocephala*	Kalimantan
Whitehead's Spiderhunter	*Arachnothera juliae*	N. Borneo
Scarlet Sunbird	*Aethopyga mystacalis*	Java
Kuhl's Sunbird	*Aethopyga eximia*	Java
Yellow-rumped Flowerpecker	*Prionochilus xanthopygius*	Kalimantan
Scarlet-headed Flowerpecker	*Dicaeum trochileum*	Sumatra, Kalimantan, Java, Bali, Lombok
Black-sided Flowerpecker	*Dicaeum monticolum*	Kalimantan
Javan Fire-breasted Flowerpecker	*Dicaeum sanguinolentum*	Java, Bali, Nusatenggara
Enggano White-eye	*Zosterops salvadorii*	Enggano
Black-capped White-eye	*Zosterops atricapillus*	Sumatra, N. Borneo
Javan White-eye	*Zosterops flava*	Kalimantan, Java

Lemon-bellied White-eye	*Zosterops chloris*	Belitung, and islands to the East
Javan Grey-fronted White-eye	*Lophozosterops javanicus*	Java, Bali
Pygmy White-eye	*Oculocincta squamifrons*	N. Borneo
Mountain Black-eye	*Chlorocharis emiliae*	Kalimantan
Java Sparrow	*Padda oryzivora*	Java, Bali (widely introduced)
Javan White-bellied Munia	*Lonchura leucogastroides*	Sumatra, Java, Bali, Lombok (introduced in Singapore)
Dusky Munia	*Lonchura fuscans*	Kalimantan

The following endemic species also occur beyond Indonesian borders (other than northern Borneo):

Black-naped Fruit Dove	*Ptilinopus melanospila*	S. Sumatra (islands), Kalimantan (islands), Java, Bali, S. Philippines
Silvery Wood Pigeon	*Columba argentina*	Small islands off Sumatra and Kalimantan, and in the Malacca Straits
Sunda Ground Thrush	*Zoothera andromedae*	Sumatra, Java, Bali, Nusatenggara, Philippines
Mountain Bush Warbler	*Cettia vulcania*	Sumatra, N. Borneo, Java, Bali, Lombok, Timor, Palawan
Sunda Serin	*Serinus estherae*	Sumatra, Java, Sulawesi, Mindanao

Further Reading

THE present book only has space to introduce the reader to the more common and representative birds of Sumatra and Kalimantan. *The Birds of Borneo* by B. E. Smythies (third edition, 1981), published by The Sabah Society and the Malayan Nature Society, is recommended for Kalimantan; it has also been reproduced as a Pocket Guide (the illustrations, and brief notes only), published by The Sabah Society and World Wildlife Fund Malaysia in 1984.

The Birds of Sumatra, An Annotated Checklist by J. G. van Marle and K. H. Voous (British Ornithologists' Union Checklist No. 10, 1988) is the definitive reference on Sumatran birds, but it does not give descriptions. All the birds of Sumatra and Kalimantan, except the endemic species listed in the Appendix above, are described in B. King, E. C. Dickinson, and M. Woodcock, *A Field Guide to the Birds of South-East Asia* (Collins, 1975), although the reader should be cautious of racial variations that occur in some species.

The sister book to the present volume, *The Birds of Java and Bali* by D. Holmes and S. Nash (Oxford University Press, 1989), describes and illustrates some additional birds not described here. These islands are also covered in *A Field Guide to the Birds of Java and Bali* by J. Mackinnon (Gadjah Mada University Press, 1988).

Index to Genera, Systematic Section